AF745272

SCIENTIFIC MANAGEMENT OF SWINE PRODUCTION AND PORK PROCESSING

SCIENTIFIC MANAGEMENT OF SWINE PRODUCTION AND PORK PROCESSING

Edited by
Dr. Jowel Debnath
M.V.Sc & Ph.D (Animal Genetics & Breeding)
Assistant Professor,
Department of Livestock Farm Complex
College of Veterinary Sciences & A.H., R.K. Nagar, West,
TRIPURA-799008 (INDIA)

NARENDRA PUBLISHING HOUSE
DELHI (INDIA)

CRC Press
Taylor & Francis Group
Boca Raton London New York

CRC Press is an imprint of the
Taylor & Francis Group, an **informa** business

First published 2026
by CRC Press
4 Park Square, Milton Park, Abingdon, Oxon, OX14 4RN

and by CRC Press
2385 NW Executive Center Drive, Suite 320, Boca Raton FL 33431

CRC Press is an imprint of Informa UK Limited

© 2026 Narendra Publishing House

The right of Jowel Debnath to be identified as the author of the editorial material, and of the authors for their individual chapters, has been asserted in accordance with sections 77 and 78 of the Copyright, Designs and Patents Act 1988.

All rights reserved. No part of this book may be reprinted or reproduced or utilised in any form or by any electronic, mechanical, or other means, now known or hereafter invented, including photocopying and recording, or in any information storage or retrieval system, without permission in writing from the publishers.

For permission to photocopy or use material electronically from this work, access www.copyright.com or contact the Copyright Clearance Center, Inc.(CCC), 222 Rosewood Drive, Danvers, MA 01923, 978-750-8400. For works that are not available on CCC please contact mpkbookspermissions@tandf. co.uk

Trademark notice: Product or corporate names may be trademarks or registered trademarks, and are used only for identification and explanation without intent to infringe.

Print and Electronic editions not for sale in South Asia (India, Sri Lanka, Nepal, Bangladesh, Pakistan and Bhutan).

British Library Cataloguing-in-Publication Data
A catalogue record for this book is available from the British Library

ISBN: 9781041247166 (hbk)
ISBN: 9781041262916 (pbk)
ISBN: 9781003749844 (ebk)

DOI: 10.1201/9781003749844

Typeset in Times New Roman
by Amrit Graphics, Shahdara, Delhi

Contents

Preface

In a country like India, livestock and its products play a vital role not only through its contribution to the Gross Domestic Product, but also due to the fact that several small farmers in rural villages are dependent either directly or indirectly on livestock rearing for their survival. Given the variety and diversity of animal genetic resources, land type, climate and other natural resources, the system of livestock production varied.

In the current scenario, with the growing human population, the demand for livestock products are ever increasing, while the available resources for animal husbandry in terms of livestock, feed, pasture, land etc. are decreasing. To meet this widening gap between availability and demand, it is important to improve efficiency of livestock production in such a way as to increase output from available resources.

Any livestock enterprise requires sound application of basic scientific principles of management to achieve maximum productivity and profitability. Scientific management of livestock requires the region and situation based comprehensive knowledge on available genetic resources, feeding, housing and disease control. Proper management at right time in proper way will support to become a successful entrepreneur or manager.

Pig farming in large scale with proper scientific management and skill has tremendous scope to boost up the rural economy of our country. But, due to lack of proper guidance or scientific farming protocol on pig rearing farmers can't ensure the proper earning in this country. Therefore, it is utmost important to make available scientific information on pig rearing for the farmers. Hence, an effort has been made to highlight on basic principles of rearing of pig, meat and meat products. The prime objective of this book is to put in the picture of pig rearing in modern scientific way to ensure maximum productivity and profitability of the farmers.

The information of this book could be of use for the B.V.Sc & A.H. students, researchers, para-vets, farmers and the persons willing to gather knowledge about swine genetic resources, swine nutrition, swing management, swine diseases and control, slaughter technique of swine, meat and meat products of swine.

As the editor of the book, I have the pleasure of expressing my sincere gratitude and appreciation to all the contributors without whose help and efforts, it would not have been possible to write this book.

I do hope and believe this book would serve as valuable practical implementation in pig farming in our country.

Jowel Debnath

List of Contributors

Dr. Jowel Debnath
Assistant Professor (S.G),
Department of Livestock Farm Complex
(Faculty of Animal Genetics & Breeding),
C.V. Sc & A.H., R.K. Nagar,
West Tripura-799008, India

Dr. Nibash Debbarma
Associate Professor
Department of Livestock Production
Management,
C.V.Sc & A.H., R.K. Nagar,
West Tripura-799008

Dr. Debasish Behera
Assistant Professor (S.G),
Department of Veterinary Pathology,
C.V.Sc & A.H., R.K. Nagar,
West Tripura-799008, India

Dr. Bikas Ch. Debnath
Assistant Professor (S.G),
Department of Animal Nutrition,
C.V.Sc & A.H., R.K. Nagar,
West Tripura-799008, India

Dr. Tapan Kumar Das
Assistant Professor (S.G),
Department of Livestock Farm Complex
(Faculty of Animal Nutrition),
C.V.Sc & A.H., R.K. Nagar,
West Tripura-799008, India

Dr. Dipanjan Chakraborty
Assistant Professor (S.S),
Department of Animal Reproduction,
Gyneacology and Obstetrics,
C.V.Sc & A.H., R.K. Nagar,
West Tripura-799008, India

Dr. Amit Kumar Nath
Assistant Professor (S.S),
Department of Teaching Veterinary
Clinical Complex(Faculty of Animal
Reproduction, Gyneacology and
Obstetrics) ,
C.V.Sc & A.H., R.K. Nagar,
West Tripura-799008, India

Dr. Mritunjay Kumar
Associate Professor, Department of
Veterinary Medicine,
Bihar Veterinary College,
Patna-800014, India

Dr. Kanika Kalai
Assistant Professor (S.G),
Department of Veterinary Pathology,
C.V.Sc & A.H., R.K. Nagar,
West Tripura-799008, India

Dr. Bijoy Kumar Sarkar
Assistant Professor (S.G),
Department of Livestock Products
Technology, C.V.Sc & A.H.,
R.K. Nagar, West Tripura-799008, India

About the Editor

Dr. Jowel Debnath is presently working as Assistant Professor (Selection Grade) in the Department of Livestock Farm Complex (Faculty of Animal Genetics & Breeding), College of Veterinary Sciences & A.H., R.K.Nagar, West Tripura-799008 since 2012. He obtained B.V.Sc & A.H degree from College of Veterinary Sciences & A.H., Selesih, Aizawl, Mizoram (India) in 2009. He did his M.V.Sc & Ph.D in Animal Genetics & Breeding from the Division of Animal Genetics, Indian Veterinary Research Institute, Izatnagar, Bareilly, U.P.-243122 in the year of 2011 and 2016, respectively. He cleared ICAR-ASRB NET in the discipline of Animal Production and Animal Genetics & Breeding in the year of 2012 & 2016, respectively and also cleared the ICAR's All India Examination, ICAR-SRF (PGS)-2011-12. He was recipient of IPSA Prof. P.K. Pani research award-2015 from Indian Poultry Science Association, ICAR-CARI, Izatnagar, Bareilly, U.P.-243122. The editor has more than 10 years of experience in the field of teaching, research and extension. He has more than 25 research papers to his credit in different reputed National and International Journals. Editor has edited and published a book namely "Duck Production & Management" in the year 2022. Editor has also published many review articles, popular articles and book chapters in various scientific journals and in books, respectively. He has also published booklets in local language for poultry farmers. Editor has organized various training for the farmers and also acted as resource person in various scientific and skilled development programmes. Editor has completed Department of Biotechnology, New Delhi, Government of India and ICAR-Central Avian Research Institute, Izatnagar, Bareilly, U.P-243122, Govt. of India funded research project as Principal Investigator and Co-Principal Investigator, respectively. His area of research is Avian Molecular Genetics.

CHAPTER 1

SWINE GENETIC RESOURCES

Dr. Jowel Debnath

1.1. INTRODUCTION

Pigs play a vital role in providing livelihood, food, and nutritional security to billions of people across the world. A profitable pig production immensely depended on two vital factors *viz* superior genetic resources and environment. Swine genetics plays an important role in meat production performance and meat quality. Apart from providing meat, it also provides fat, bristles and manure. At present, there are 137 recognized pig breeds in the world.

1.2. POPULATION STATUS

India's pig population is 9.06 million (20th Livestock Census, 2019), against the world population of about 900 million (FAOSTAT Databases 2020). About 79.03% of the total pig population is of indigenous & nondescript pigs (20th Livestock Census, 2019).

1.3. INDIGENOUS SWINE BREEDS OF INDIA

There are 14 registered indigenous swine breeds in India. Indigenous pigs are mostly small in size, pot-bellied with a prolonged face and short ears. The indigenous pig breeds are as follows:-

1. **Ghungroo**: Ghungroo is the first recognized pig breed of India (Accession No: INDIA_PIG_2100_ GHOONGROO_09001). This pig is found in North Bengal and the adjoining district of Assam.

DOI: 10.1201/9781003749844-1

Utility: Pork

Colour: Mostly black coloured, occasionally hooves and metapodials are white.

Characteristics: Thick coarse and long hair coat, long tail, upwardly curved snout, broad and flattened face with large and heart shaped ears, long and cylindrical barrel belly and seven pairs of thoraco-abdominal teats. This pig breed is highly prolific with a litter size of 6-10 piglets. The mean body weight at birth is about 0.89-1.1 kg and adult body weight is about 55 to 60 kg. This breed can be reared in a low input production system. The female and male pigs attain sexual maturity at 10 months and 8 months, respectively. The gestation period is about 112-115 days.

2. **Agonda Goan:** Agonda Goan is a pig breed (Accession No: INDIA_PIG_3500_ AGONDAGOAN _09003) found in the Agonda region of Goa. This pig is commonly known as 'Gavthi Dukor' in Goa.

   Utility: Pork

   Colour: Mostly black, few animals (both male and female) have white patches on legs and face.

   Characteristics: These are small sized having short ears, short snout, pot belly and are wild in nature with rough bristles. The mean weight at birth and weaning is about 0.59±0.01 kg and 3.76±0.12 kg, respectively. The yearling weight of adult male and female is about 46.58±4.50 and 47.27±6.46 kg, respectively. The average litter size at birth and at weaning is about 7.21±0.41 and 5.93±0.28, respectively. Pre-weaning and post-weaning growth rates are about 79.25±9.86 and 153.85±13.34 g/day, respectively. Average age at slaughter is about 6-8 months in male with a live weight of about 67kg. Females are good mothers, they take care of their piglets very well due to which there is less piglet mortality during early period by stampeding. The unique trait of Agonda Goan pig is good maternal instinct and lean meat with high palatability. The average age at first estrus, age at first conception and age at first farrowing are 172.4±22.95, 206±19.26 and 324.75±28.15 days, respectively.

3. **Niang Megha:** Niang Megha is a medium-sized pig breed (Accession No: INDIA_PIG_1300_NIANGMEGHA_09002) found in the Garo, Khasi, and Jaintia hills of Meghalaya.

   Utility: Pork and bristle

   Colour: Black, star shaped white patches at forehead and sometimes hock joint with erect bristles on the dorsal midline.

   Characteristics: Snout-medium tapering partially white at nostril, bristle-long on midline and uniformly distributed in other places of the body, hoof placement partial, top line-straight in male, concave in female, 4-6 pairs of mid ventral teats. The adult average body weight ranges between 50 to 60 kg. The averages litter size at birth, litter size at weaning, litter weight at birth, litter weight at weaning, age at first estrus, age at the first fertile service, age at first farrowing, farrowing interval and the gestation period is 6.080 ± 0.219, 5.202 ± 0.190, 3.172 ± 0.107 kg, 30.614 ± 1.199 kg, 221.173 ± 1.527 days, 246.440 ± 1.385 days, 347.813 ± 3.516 days, 206.121 ± 0.785 days and 111.848 ± 0.136 days, respectively.

4. **Tenyi-vo:** Tenyi-vo pig is a registered pig breed (Accession No.: INDIA_PIG_1400_TENYI-VO_09004) native to the state of Nagaland.

   Utility: Pork and bristle

   Colour: Mostly black in both sexes. The body coat is predominantly jetblack with white markings on the forehead, flanks, switch of the tail and legs.

   Characteristics: These are potbellied animals with sagging back and pendulous belly touching the ground in females, straight tail ending with white switch reaching the hock joint, white stocking, white markings on the forehead and ventral body. Tenyi- Vo pigs have strong and long tapering snout, small erect ears and bright alert eyes. The adult body weight ranges between 40 to 60 kg at maturity. The female and male pigs attain sexual maturity as early as 150 days and 90 days, respectively. The gestation period is 109-116 days and usually farrow 4-6 piglets.

5. **Nicobari:** Nicobari is a pig breed (accession No. of the breed is INDIA_PIG_3300_ NICOBARI _09005) found in Car Nicobar and adjoining islands of Andaman and Nicobar. This pig is locally called as Ha-un and closely linked with the local culture and tradition.

   Utility: Pork

   ***Colour*:** Mostly black to red-brown, brown, grey, and fawn. Few are creamy-white, reddish-brown and black and brown mixed.

   ***Characteristics*:** Sturdy and short with long body. Medium to short snout and short erect ears. Marked bristle crest (mane) on the back extending from mid head/shoulder to base of the tail. Facial profile varied from flat to concave. Neck is short with very large jowl. No curling is the characteristic feature of the tail. The average adult body weight is 45-55 kg. The average age at first farrowing, litter size and the farrowing interval is 10 to 12 months, 6 to 10, and 8-10 months, respectively.

6. **Doom:** The Doom pig is the only registered pig breed of Assam (Accession No INDIA_PIG_0200_ DOOM _09006).

   Utility: Pork and bristle

   ***Colour*:** Predominantly black with a white patch on the forehead.

   ***Characteristics*:** These pigs have short concave snout. They are large; flat belly type with short erect ears. Top line is straight with long bristles extending up to thoraco-lumber area. The average adult body weight of doom pig is about 35 to 45 kg. The average age at first estrus, age at first fertile service, age at first farrowing, gestation length, and inter farrowing period is 213.91±1.36, 237.95±1.18, 351.51±1.24, 113.51±0.40, and 213.25±1.45 days, respectively. The average litter size at birth, litter size at weaning, litter weight at birth, and litter weight at weaning is 5.55 ± 0.11, 4.56 ± 0.09, 3.38 ± 0.06 kg and38.75 ± 0.55 kg, respectively.

7. **Zovawk:** Zovawk pig is a registered pig breed (Accession No.: INDIA_PIG_2700_ ZOVAWK _09007), native to the state of Mizoram.

   ***Utility*:** Pork

***Colour*:** Predominantly black with a star on the forehead and white hairs in four limbs below the knee region.

Characteristics: Erect ears, concave snout, pot belly, concave top line and long bristles on mid-line. The adult body weight ranges from 40 to 50 kg at maturity. The female and male pigs attain sexual maturity at 120±1.18 days and 143.20±1.44 days, respectively. The average age at first farrowing, gestation period is 314.9±5.8 days and 114.77±0.27 days, respectively and usually farrow 5-8 piglets.

8. **Ghurrah:** Gurrah pig is (accession No. of the breed is INDIA_PIG_2000_GHURRAH_09008) native to Bareilly division and adjoining pats of Lucknow division of Uttar Pradesh.

   Utility: Pork

   ***Colour*:** Mostly black with a white stripe below the hock joint. Few are greyish black or brownish.

   ***Characteristics*:** Medium sized pigs with flat belly, angular body and long straight snout. Legs below hock joint are white. Thick line of hairs is present from neck to shoulders. Head is elongated with triangular face and short leaf shaped vertically erected ears. The adult male body weighs is about 46kg and the female is about 48kg. The average litter size is 6.85 at birth and 5.65 at weaning.

9. **Mali:** Mali is an indigenous pig breed (Accession No: INDIA_PIG_1900_MALI_09009) native to Tripura.

   Utility: Pork

   Colour: The pig is predominately black coat colour and has no specific distinct marking.

   Characteristics: A medium to a small bristle is ubiquitously distributed throughout the body. They are medium-sized with an adult body weight of about 68 kg and 71 kg in males and in females, respectively. The female and male pigs attain puberty at 117.9±1.5 days and 138.3±6.4 days, respectively. The average age at first farrowing is about 281.4±1.6 days (Dandapat *et al.* 2010). The average litter size is 5.15 (range 3-7) at birth and 4.46 (range 3-6) at weaning age.

10. **Purnea Local:** Purnea Local pig breed (accession No. of the breed is INDIA_PIG_0325_ PURNEA_09010) is native to the Purnea and Araria districts of Bihar but is also found in other parts of Bihar and some parts of Jharkhand.

    Utility: Pork

    Colour: Predominately black in colour.

    Characteristics: Purnea Local pigs are mainly characterized by round faces, short erect ears with a slightly concave snout. These pigs are with potbelly but there is no specific distinct marking. However, in some pigs, a white spot at the lower limb is also seen. The thick line of bristle is present from neck to shoulders. They are medium-sized with an adult body weight of about 44 kg and 42 kg in males and females, respectively. The average litter size at birth and weaning is 5.02 and 4.97, respectively.

11. **Banda:** Banda pig is native of Jharkhand (accession No. of the breed is INDIA_PIG_2500_BANDA_09011)

    Utility: Pork

    Colour: Predominantly black in colour.

    Characteristics: Banda local pig is mainly characterized by black colour with short and erect ear. These animals are having medium to short bristle on neck with a long and concave snout. These are potbellied and characterized with small litter size. Adult body weight averages 28.0 kg in males and 27.0 kg in females. Average litter size is 4.5 (range 4-7) at birth and 4.25(range 4-6) at weaning.

12. **Manipuri Black:** Manipuri Black pig is native of Manipur state (accession INDIA_PIG_1200_MANIPURIBLACK_09012)

    Utility: Pork

    Colour: Black in colour

    Characteristics: Manipur Black pig is medium in size with flat belly and short legs. The head of these pigs is short, slightly concave with short ears, and short to medium snout. White patches are sometimes seen in extremities such as legs and snout area. Hairs are predominantly black and sparse, however few pigs with dark grey are also found. Bristle production is very scanty and cutting has not been practiced. Adult body weight averages about 96.0 kg in males and 93.0 kg in females. Average litter size is 8.27 (range 6-11) at birth and 6.02 (range 5-9) at weaning.

13. **Wak Chambil:** Wak Chambil is mainly distributed in Garo Hills of Meghalaya state (accession INDIA_PIG_1300_WAKCHAMBIL_09013)

    Utility: Pork and its unique flavour and taste and cherished during religious and ceremonial occasions by local people.

    Colour: Black in colour

    Characteristics: Wak Chambil is a small sized pig with round and pendulous belly. Average adult body weight is 32.0 kg in males. Litter size at birth ranges from 4 to11.

14. **Andamani Pig:** Andaman local pig, also known as Andamani pig, is an indigenous pig germplasm of Andaman and Nicobar Islands and is distributed in different islands of Andaman group of islands (INDIA_PIG_3300_ANDAMANI_09014).

    Utility: Pork

    Colour: They are sturdy and medium in size and black (mostly) or brownish in colour.

    Characteristics: Neck and back portion hairs are very thick as well as long whereas hairs on the sides and flank regions are shorter and thinner relatively. The most commonly observed feature is slightly downward arch/ curvature of the back (low back). They attain body weight of 60-75 kg for male and 55-65 kg for female at 1 year of age under field condition. Height ranges from 65 to 89 cm. Average adult body length for male and female are 115.80 ± 2.05 cm and 104.60 ± 1.28 cm respectively. Snout is short to medium with slight concave type. Ears are short and erected upwards (no hanging ears). Litter size at farrowing ranges from 6-13.

1.4. EXOTIC SWINE BREEDS

Classification of exotic swine breeds

1. **Lard Type:** Lard pigs which tend to be short of body length, grow slowly, not as large and put on a lot of fat easily. E.g. Poland China and Chester weight
2. **Bacon type:** Bacon pigs sometimes also called as meat pigs which have longer bodies and develop higher levels of muscling. These breeds are generally faster growing and larger. In comparison to lardpigs, the diameter of the loin muscles is well developed and much larger. e.g. Large White Yorkshire, Landrace, Tamworth, Berkshire, Hampshire and Duroc.

Breeds:

1. **Large white Yorkshire:**

   Origin: Yorkshire and neighbouring countries of northern England.

   Utility: Bacon

   Colour: The body colour is solid white.

   Characteristics: Yorkshire sows are known as good mothers. Other distinguishing characteristics are erect ear, snout of medium length, large in size and slightly dished face. Skin is pinkish in colour and wrinkles free with long and moderately fine coat. Neck is long and full to the shoulder. Back is slightly arched. Matured body weight of boar and sow is 300-400 kg and 230-320 kg, respectively. Excellent breed for crossbreeding and most extensively used in India.

2. **Landrace:**

   Origin: Denmark

   Utility: Bacon

   Colour: White in colour

   Characteristics: Large in size, long body with relatively short legs and heavy lop ears. White bristles in white body. Matured body weight of boar and sow is 270-360 kg and 200-320 kg, respectively.

3. **Duroc**

   Origin: Denmark

Utility: Bacon

Colour: The breed is moderately red in colour with shades varying from a golden to cherry red in colour.

Characteristics: Ducroc is excellent for weight gain rate and high feed converter. The sow matures early, produces large litter (up to 15 piglets) and fair mothering ability. Matured body weight of boar and sow is 400 kg and 350 kg, respectively.

4. **Poland China**

   Origin: China

   Utility: Bacon

   Colour: Black in colour with white patches on face, feet and snout.

   Characteristics: It has large body size and floppy ears. It has strong bone and well-developed legs and feet. It is prolific (produce 16-17 piglets per farrow) and good meat and carcass quality. A mature boar weighs between 250kg and 360 kg and mature sow weighs between 225 kg and 300 kg.

5. **Middle White Yorkshire**

   Origin: England

   Utility: Pork

   Colour: White

   Characteristics: Short head, upturned dished face, short snout. Neck blend from head to shoulder. Legs straight and short, erect ears. Adult body weight boar: 240-340 kg and sow: 180-270 kg.

6. **Berkshire:**

   Origin: England

   Utility: Pork

   Colour: Black colour with six white points on feet, nose and tail.

   Characteristics: Medium size. Short head with dished face, mature body weight of boar and sow is 270-380kg and 200-290kg, respectively.

7. **Tamworth:**

   Origin: USA (New york and new Jersey)

Utility: Pork

Colour: Cherry red from light to rich reddish brown.

Characteristics: Ears 2/3rd lop, slightly concave nose, arched back. Popular for size, hardiness, prolificacy. Sows are excellent mother. Mature body weight of boar and sow is 400 kg. and 350kg., respectively.

8. **Chester White**:

Origin: Chester & Della, USA

Utility: Lard

Colour: White hair and skin

Characteristics: Ears are drooping. Hardy fairy good feeder. Prolific exceptional mothers. Mature average body weight is 400 kg.

9. **Wessex Saddleback**:

Origin: Britain

Utility: Bacon

Colour: The Wessex Saddleback is a striking looking black pig with a white belt, which includes the front legs, around the body.

Characteristics: The Wessex is both prolific and hardy. The ears are lopped forward. Mature average body weight of boar and sow is 400 kg.

CHAPTER 2

ECONOMIC TRAITS IN SWINE

Dr. Jowel Debnath

2.1. ECONOMIC TRAITS

Traits those influence the economy of the farm are called as economic traits.

2.2. FACTORS INFLUENCING ECONOMIC TRAITS

Phenotype	:	Genotype +Environment
Genotype	:	Species, Breed, strain, variety
Environment	:	Macroclimatic conditions: Herd, year and season
		Microclimatic conditions: Temperature, humidity and other managemental conditions.

2.3. VARIOUS ECONOMIC TRAITS OF SWINE

1) **Litter size at birth**: Number of piglets born in a litter.
2) **Litter size born alive**: Number of piglets born alive in a litter.
3) **Litter size at weaning**: Number of piglets weaned per litter at 56th day.
4) **Birth weight**: Weight of a piglet in kg at birth
5) **Litter weight at birth**: Weight of all piglets born alive of a litter in kg at birth.
6) **Weaning weight**: Weight of a piglet in kg at 56 days of age.

DOI: 10.1201/9781003749844-2

7) **Litter weight at weaning**: Weight of all piglets of a litter in kg at weaning.

8) **Weight at market age**: Weight of a pig in kg at the market age of 154 or 210 days.

9) **Growth rate**

   a) **Pre-weaning growth rate**: Daily weight gain in kg from the date of birth to date of weaning.

$$\frac{\text{Weaning weight } (56^{\text{th}}\text{day}) - \text{Birth weight}}{56}$$

   b) **Post-weaning growth rate**: Daily weight gain in kg from the date of weaning to date of marketing.

$$\frac{\text{Weight at market (154 or 210 days)} - \text{Weaning weight}}{\text{Age at market (154 or 210 days)} - 56}$$

10) **Feed efficiency**

   a) **Weaning to 154 days**: Weight of feed consumed in kg for producing one kg of live weight from weaning to 154 days.

   b) **Weaning to 210 days**: Weight of feed consumed in kg for producing one kg of live weight from weaning to 210 days.

11) **Mortality percentage/EADR**

   a) **Pre-weaning mortality**: Number of piglets died from birth to 56th day (weaning) to the total number of piglets born alive, expressed in percentage (10 to 15%).

   b) **Post-weaning mortality**: Number of piglets died from weaning to 365 days to the total number of piglets weaned, expressed in percentage (5 percent).

   c) **Adult mortality**: Number of adult pigs died during the year to the number at the beginning of the year, expressed in percentage.

Average values:

Traits	Large White Yorkshire
Birth weight (kg)	1.25-1.70
Weaning weight (kg)	8.50-9.00
8 months weight (kg)	90 -100
Litter size at birth	9-12
Litter size at weaning	8-9
Litter weight at birth (kg)	13.5-16.5
Litter weight weaning (kg)	72-81
ADG – pre-weaning (g/day)	160
ADG – post- weaning (g/day)	360-375
Pre-weaning mortality (%)	10
Post- weaning mortality (%)	<5
Adult mortality (%)	<5

CHAPTER 3

SWINE PRODUCTION AND MANAGEMENT

Dr. Nibash Debbarma and Dr. Debasish Behera

3.1 GENERAL MANAGEMENT OF SWINE

Pig farming is a crucial aspect of agriculture as it provides abundant benefits to farmers. Raising pigs can be highly profitable and rewarding for beginners in the farming industry because the cost of production is low compared to other livestock farming businesses. It requires low to medium capital to start. Therefore, the pig farming plays a vital role in the agricultural sector, contributing to food security and economic growth. Pigs are excellent converters of feed into high-quality protein, making them a valuable source of meat. Raising pigs can be a lucrative venture, as pork consumption is in high demand globally. The market for pork products and by-products is vast, ensuring a profitable return on investment. Pigs also have a high disease resistance and adopt easily to most environments. Because of this adaptability pigs make excellent candidate for intensive or diversified agriculture that fits a varied range of budgets.

Other than meat, it also provides bristles and high valued organic manure and has been established as a popular farming due to its inherent characteristics of faster multiplicaion, higher growth rate and feed conversion ability. Pig rearing in India is carried out under diverse social, climatic and environmental conditions. Pigs are more acceptable to rural population contributing substantially to the rural well being and self sufficiency due to the better returns and suitable backyard venture. The revenues earned from pork production have been limited due to the lack

DOI: 10.1201/9781003749844-3

of input intervention of modern technology especially in the rural areas. Significant improvements in swine production can further be made by adopting best possible management practices. In the areas where pigs are reared on free range, they are most valued as a kind of "savings" to the farmer so as to encash in times of financial emergency. Commercial production under semi-intensive and intensive conditions are becoming more popular because of its high demand and less supply as well as favourable return on investments.

3.2. SCOPE OF PIG FARMING

1. Pig requires comparatively small investment on housing and equipment compared to other livestock.
2. Pig farming is well adapted to both diversified and intensive agriculture in rural and semi-urban areas.
3. Pigs are being considered as efficient converter of agricultural byproducts and waste material into high quality protein.
4. It is the only litter bearing animal among meat producing livestock having shortest gestation period of 111 days.
5. They are prolific breeders with 2 farrowings every 1 ½ years and commonly farrowed from 8 – 12 piglets.
6. The pork is an important source of high-quality protein and carcass return is quite high as 65 – 80 % of live weight.
7. The bristles are used in making brushes and the faeces of pig are useful in maintaining soil fertility and feed for fishponds.
8. Pork is most prefererred in terms of delicacy due to higher content of fat and the lower content of water, the energy value of pork is higher than that of beef or lamb.
9. Gilts may be bred at 8 months of age and the pigs can be marketed at 6 months of age after farrowing.
10. There is good demand from domestic as well as export market for pig products such as pork, bacon, ham, sausages, lard etc.

11. Pig store fat rapidly for which there is an increasing demand from poultry feed, soap, paint and other chemical and leather industry.
12. The pig has 15,000 taste buds as against 9,000 in humans enabling it to eat everything edible to man and other animals including forage.
13. The pig can tolerate a higher level of dietary fibre than has been previously recognised has greatly expanded the range of feed resources for pigs.
14. Pigs are the most efficient animals for converting kitchen wastes and other non-conventional feedstuffs into meat.
15. The pig is a more efficient carcass yielder than cattle, sheep or goats, dressing out at about 70% compared to between 50-55% for cattle and about 50% for sheep and goats. In addition, pig carcasses have a smaller proportion of bones and higher proportion of edible meat.
16. Pig carcasses are easier to dress and have superior curing qualities, an obvious advantage for processing and marketing.
17. The pig possesses a large ceacum, the dropping from pigs is rich in nutrients. Pig faeces make good fertilizer for crops and can also be recycled into livestock feeds.
18. In comparison to dairy and poultry farming, pig farming requires very less initial capital investment and additionally rate of return is very high.
19. Pig husbandry does not require much scientific and technical input so, can be easily adopted by poor and less literate peoples.
20. Due to high growth rate and high prolificacy, pig produces much more live weight in specified time from per animal. Therefore, pig is an excellent meat animal.
21. Feed is most important factor affecting economic feasibility of any livestock husbandry activity. Pig is an efficient converter of agriculture by products, kitchen waste and other products.
22. Pig needs very simple housing structure and less labour intensive, so, women and old age person can be utilized as household labour.

3.3. PIG PRODUCTION SYSTEMS

Systems of pig production commonly classified into

1. Small scale system
2. Large scale system
3. Integrated system

1. **Small Scale System**

   Scavenging System: This is a traditional system of rearing pigs in large part of the tropics and considered as the cheapest system. The pigs are allowed to wander freely and pick up food but low nutritional quality especially banana, rice bran, local herbaceous plants and kitchen waste. Indigenous pigs are predominating on the scavenging system, because they are adapted to local environment. The productivity of pig is normally low with erratic feed supply resulting in irregular breeding of sows, high rates of piglet mortality and low growth rates. But the major disadvantages of scavenging system are that pigs have access to source of parasites, human excreta, other infectious diseases.

   Semi-intensive System: This type of system is also known as backyard system. The pigs are confined in rearing pens and sties are mainly of simple construction with reasonable cost. In this system, sometimes pigs are constrained by tethering in larger yards or paddocks.

2. **Large Scale System**

   Intensive System: It is the most common system of large-scale production. This system is generally capital intensive and with more number of animals. Modern high-performance breeds of pigs are used and provided with optimum conditions of housing, feeding and management in order to ensure maximum output.

   Extensive Systems: This system is also known as outdoor system, keeping sows in paddocks and providing individual huts for farrowing and shelter. These types of system commonly exist in the tropics. Advantage of this system is less capital investment and sows can gain access to bulky foods such as pastures and crop residues. Moreover, it is essential that adequate shade and wallows are provided and parasite infestation should be controlled with proper feeding of anthelmintic.

3. **Integrated System**

   Integration of pig production with other enterprises and other farming systems are being widely practiced in the tropics, involving various combinations of fish farming, production of algae, methane gas production, ducks, water hyacinth and vegetable production. Such types procedure enhances the efficiency of resource use and increase output for overall operation.

   Example: Fish ponds with sufficient water and pig sties can be constructed near by and required amount of manure/excreta can be utilized as a biofertilizer for better growth plankton and phytoplankton and that can be the source of feed for the optimum fish production.

3.4. MANAGEMENT FOR OPTIMAL PRODUCTION

The fertility of sow and litter size farrowed and weaned are very important factors in determining the viability of swine projects. The cost of raising 6 pigs or 10 pigs from birth to weaning is more or less similar since the maintenance requirement of the sows is the same to both cases. Breeding and selection alone are not very certain methods of improving the production of swine. Time spent on improving the management of the sow and litter, on increasing the efficiency of feed utilization and the control of diseases may contribute more to the income of the enterprise than an equal time spent to increase herd productivity by following complex method of breeding and selection. To ensure that, the maximum number of bred gilts and sows farrow large litter of living pigs, the following measures are suggested.

i. Gilts should weigh at least 100 kg before breeding. Gilt may be bred to farrow at 12-14 months of age. Gilts come into heat during first few days after weaning and after 3 weeks it can be bred on the first heat.

ii. In hand breeding sows are bred twice during the heat period and again after 24 hours. In pen mating sow may be put with a boar till the heat lasts.

iii. The commercial pig may be bred to several boars in the same heat period. Two boars are run together with the same female in heat or one boar for 1 day and second for next. With this procedure the chance of conception will increase.

iv. Though the heritability of litter size is low and therefore, selecting for this trait should be continued. In order to increase the litter size by 1 pig we may have to select pigs for this trait for 20 generations.

v. The female selected for breeding must have at least 12 functional and well spaced teats and a good body length. While selecting the breed for an enterprise the litter size must be kept in mind. Yorkshire pigs are on the top in this regard with 10-11 pigs per litter. In most breeds litter size is 9-10 but, in few breeds, it may be as low as 6-8.

vi. If old boars are to be used for breeding, one with a good conception rate with large litters may be used.

vii. Both the gilts and sows must be thrifty and in medium condition before breeding, if it is too fat, the condition may be brought down to proper weight and if too thin, the health condition may be improved by proper management.

viii. During hot summer months one must keep the breeding and pregnant sows cool by water sprinkling or by the use of wallow.

ix. Intense inbreeding may be avoided as for each 10% inbreeding there is one-third less pig per litter farrowed and one-half pig less at weaning.

x. Before breeding check the females for brucellosis and leptospirosis. One must avoid breeding sows suffering from flu. During early gestation high body temperature at breeding time may cause abortion.

xi. Gilts and sows may be kept in the farrowing stall or pen 5-7 days before due to farrow.

xii. A well-trained attendant is needed for farrowing. As soon as newborn pigs are dried after removing membranes, it is made sure they breath. Overlying should be prevented. Greatest losses occur within 3 days after farrowing, so a close watch on the sows and pigs must be maintained during that period. The death may be due to starvation (26%), digestive problem (2%) and unknown cause (13%).

The sow is at her best in milk production at about 3 weeks lactation after which milk quantity is reduced. So, a creep feed may be started from 3 weeks till weaning to supplement milk. The ideal weaning weight at 8 weeks age is 15 kg. The farrowing and nursing facilities between groups of females farrowed must be cleaned and kept vacant for at least a week. The sow and pigs must be put on 'clean pasture'. More than 6 sow and litter per lot may not be put on the pasture.

3.5. AGE OF BREEDING STOCK

1. Well-developed gilts may be bred to farrow when 12 to 14 months old. Fertility of an individual is affected by age.
2. Gilts should weigh at least 100 kg before breeding. Ovulation rate increases during successive oestrous periods (up to fifth) following puberty. Thus, it is advantageous to delay breeding of gilts until the second or third oestrus.
3. Litter size increases, on an average, in succeeding pregnancies up to fifth or sixth litter.
4. It is, therefore, advantageous to cull the sow from a breeding herd or a commercial herd after her fifth or sixth litter, as the litter size goes down thereafter.

3.6. SELECTION OF A BREED

1. Hampshire breed excel in carcass quality whereas Yorkshire do so in prolificacy. Most of the exotic breeds raised in the temporate climates do well in tropical and sub tropical climates. But, no particular breed of pig can excel in all points of swine production. Each breed has its own peculiarities, advantages and disadvatages in terms of feeding, breeding, management and production.
2. Middle and Large White Yorkshire do very well under Indian condition and so Durocs and Hampshire.
3. White colour being most preferred one. Middle white Yorkshire, Large White Yorkshire and Landrace are most commonly used in India.

3.7. BREEDING SYSTEMS

The following procedures should be adopted for breed improvement.

1. **Selection and Out-crossing**: A common method of breeding and multiplying purebred swine. This is being adopted in all pig-breeding farms and by farmers those produce breeding stock. The best procedure is intensive selection and out-crossing (mating of unrelated individuals of the same breed). Generally, the results are quite good as some vigour will be obtained in the offspring with some gain in litter size, livability and growth rate. Out crossing is also done frequently to introduce a trait, which is lacking in a herd.
2. **Grading Up**: The non-descript indigenous pig forms the majority of the pig population in the country. It would be advantageous to grade up this population by the successive use of boars of either Yorkshire or Landrace breed, depending on the suitability of the breed to be used in the area or localities as because of climatic conditions, availability of feeds and demand in the local markets. This would bring about an improvement in the productive traits of indigenous pigs in graded stages. This method is also appropriate to several areas when management and feeding conditions do not justify the introduction of high-quality purebred stock.
3. **Cross Breeding**: In a number of areas, particularly in the neighbourhood of bacon factories, pig production with purebred stock of Yorkshire or Landrace breed has been introduced. It would be advantageous in these areas to introduce crossbreeding, e.g. crossing Yorkshire sows with a Landrace boar and *vice versa.* While there are several systems of crossbreeding, under the prevailing conditions either single crossing or criss-crossing would be suitable and can be practiced in intensive farming system.
4. **Single Crossing**: Single crossing is the mating of a boar and a sow of two different breeds. The resulting crossbred pigs are all marketed and the cross is repeated for the next crop of pigs. This system has one disadvantage in that, the sows as well as boars have to be purchased and brought into the herd and this may be the one way of introduction of diseases into the stock.

5. **Criss-crossing**: It is the alternating use of boars of two breeds on the female stock produced in a herd. For example, Yorkshire gilts are bred to a Landrace boar. The resulting crossbred gilts are mated to a Yorkshire boar. The next cross is bred back to a Landrace boar. The gilts and sows, which are sired by a boar of one breed, are always mated to the boar of the second breed.
6. Crossbreeding can increase litter size (crossbred sows farrow and wean larger litters.), livability (greater resistance to environmental stress) and rate of growth.

3.8. SYSTEMS OF CROSS BREEDING

1. **Two – Breed Rotational Cross:** This system is often called the back cross or crisscross method. Boars of two different breeds are used in alternate generations. Crossbred gilts are retained and mated to boars from one of the two-parent breeds. The two breeds should complement to each other eg. litter size, growth and carcass characteristic etc.
2. **Three – Breed Rotational Cross:** The three – breed rotational cross is most widely used breeding system. It increases the hybrid vigour by use of three breeds instead of two. A third breed can reinforce certain performances characteristic of the crossbred offspring.

3.9. SYSTEMS OF FARROWING

1. **One Litter in a Year**

   Advantages: In this system gilts are used and pigs are farrowed in spring season. As soon as first litters are weaned the gilts become ready to market. Thus, the return value of female is more in this system. Other advantages of this system of farrowing are as follows.

   i. The management problem such as feeding, breeding etc., are less.

   ii. Less capital is required for the construction of sties, equipment and less grain storage facilities.

   iii. The weather is more favourable at the farrowing time and hence loss of piglets just after farrowing is less.

iv. The labour requirement is also not as much of other system of farrowing

Disadvantages: The optimum use of available buildings and equipment is not made. The improvement is not as rapid as in the two – litter system since the females are after each farrowing. Moreover, the distribution of labour and farm income is not uniform throughout the year.

2. **Two Litter a Year System**

Advantages: In the both fall and spring seasons, the pigs are being raised from the same sow and therefore, the pigs can be marketed at two different times of the year. The capital investment in building and equipments are being made at its best possible way. There is also better distribution of labour and farm income.

Disadvantages: The cost of maintaining dried sows throughout the year is high. By the time their useful life is over and their market value may decrease. The pig losses in two litter in a year is more.

3. **Multiple Farrowing System**

This programme can increase the size of the enterprise. The sow herd is split into 2 or 3 groups with each group farrowing twice or thrice in a year. With these groups the farrowing schedule may be as follows.

First group: January and July

Second group: March and September

Third group: May and November

In this system there is regular flow of pigs to the market, and regular flow of income. Besides that, a better use of existing facilities and distribution of labour evenly throughout the year. Two main disadvantages of the system are possibility of frequent disease outbreak and necessity to a competent help or assistant for prolonged farrowing.

3.10. BREEDING MANAGEMENT

i. The important characteristics such as size of litters, strength, vigour of-litters, milking ability, temperament, weight gain and feed efficiency of the progeny, longevity, fertility and freedom from defects are need to be considered in developing a good sow herd.

ii. Few breeds do excel in the various productive traits but studies show no breed is superior or better in all economical traits.

iii. The selection of individual animals from a herd is more important than the selection of a particular breed.

iv. Each producer at the time of establishing the herd should purchase animals from a reliable disease-free herd and should obtain all information about the animals such as feeding behaviour, history of any diseases and vaccination etc.

v. The selection of gilts and boars for replacement in the breeding herd should be based on the type and the performances of animals such as selection of gilts, culling of sows and selection of boars.

3.11. SELECTION OF GILTS

i. The gilt weigh about 90 kg should be selected for the breeding.

ii. The gilts to be selected should have been kept under similar feeding and management conditions so that any differences can be recorded as hereditary differences.

iii. The gilts performance will be a better indication of what to expect from their offspring, if it is exposed to similar management system as we do provide to other animals proposed for market.

iv. By providing special feed, care and management, the breeding worth of gilts cannot be assessed.

v. Selected gilt should be from sows, which have consistently farrowed, and weaned large litters, have reached market weight in minimum time, and have desirable market type characteristics and performances in terms of production and demand.

vi. Besides that, it would be necessary to choose gilts whose litter mates and other full-sibs have given good performance in daily gain in body weight and feed-conversion efficiency.

3.12. SELECTION OF BOARS

i. In a commercial enterprise, such as small piggery units maintained by farmers in the villages, the boar will probably be purchased.

ii. In a swine-breeding station, with adequate sow strength, it should be possible to maintain separate boar lines.

iii. The performance of the herd boar to be selected based upon high standard and it should be purchased from a breeder or a farm maintaining adequate information on its performance.

iv. The herd boar should be from a dam, which has consistently farrowed and weaned large litters. A good boar must weigh 90 kg in about 6 to 7 months, and will be strong on feet and legs.

v. Efficient feed utilization, although not mentioned in the selection of gilts, is a valuable aid in appraising a prospective herd boar.

vi. The boar's feed-conversion record from weaning to 90 kg weight would be the most desirable, apart from that, the feed conversion of tested litter mate full sibs or the tested progenies of both sire and dam would be a rational second-best technique for the selection of the herd boar.

vii. The selection of a rapidly growing boar should improve the efficiency of feed utilization as because the growth rate and efficient feed utilization are highly correlated.

Some of the Standards to be observed while replacing boars and gilts are as follows:

i. The mother to be selected should have 8 piglets or more. The minimum 56 days weaning weight of a litter (10-12 kg), in case of gilt should be 120 kg and the sow should not be less than 150 kg.

ii. The gilt or the boar should have reached a body weight of about 90 kg in about six to seven months.

iii. The pig should have adequate length and depth of the body, thick, well-muscled hams, and should be firm and trim; it should have a prominent neck, sound feet and legs with good bone.

The back-fat probe in live animals is not yet being practiced in country like India. But the following standards may be followed:

Gilts - 4 cm or less; boars, 3-2 cm or less. In addition, Gilts should have a minimum of 12 evenly spaced, functional teats. An animal with blind teats should be avoided as there will be little or no milk from these teats and the defect is heritable. Blood tests for brucellosis and leptospirosis should be made, and the animals must be immunized against swine fever. Apart from that, the pigs should be free from other diseases and physical defects.

3.13. CULLING OF SOWS AND BOARS

Culling of Sows

i. One third of the older sows annually (33% of sows).
ii. Gilts or sows not settled after 4 consecutive breeding period.
iii. Nervous and irritable nature.
iv. Produce small size litter.
v. Sows with defective teats and poor milkers.
vi. Sows with small vulva, because it is considered as indication of internal reproductive defect.
vii. Sows/gilts with inverted teats.
viii. Gilts and sows which do not meet the best standard of meat.

Culling of Boars

i. Infertile one (lack of libido)
ii. Over 5 years of age
iii. Ill-tempered nature and nervous disposition
iv. Obese and too heavy, finds difficult to mount
v. Cryptorchid condition
vi. Weak limbs

3.14. MANAGEMENT AT BREEDING OF PIGS

Feeding gilts and dry sows liberally to increase energy intake 10 to 15 days prior to mating is called *flushing*. It may be done as follows:

i. Feed leguminous hay (Cowpea / lucerne / berseem) inorder to supply more protein, minerals and vitamins.
ii. Extra allowance of grains to be provided.
iii. Administer of multivitamin injection along with "flushing".

Advantages of Flushing

i. Physical wellbeing or health of female is maintained.
ii. Prompt post weaning oestrus and prominent heat symptoms.
iii. Increases ovulation rate.
iv. It gives good litter size and high numbers born.
v. Shortens the period between weaning to successful conception.

3.15. FLUSHING AT BREEDING

i. Optimum attention towards the feeding of sows and gilts before breeding.
ii. 'Flushing' (feeding liberally before breeding) to sows and gilts at breeding time increase their ovulation rate.
iii. A decent grower ration to be fed to sows and gilts seven to ten days before breeding is excellent for flushing.
iv. After breeding, sows and gilts should be fed a well-balanced ration until the last six weeks of pregnancy.

3.16. FLUSHING OF GILTS

It is the practice of increasing the energy intake in gilts for 7 to 10 days (1 – 2 weeks) before breeding or at about 7 ½ months of age. They should be fed to 6 to 8 lbs (2 ½ - 3 ½ kg) / day to provide the flushing effect.

Advantages

i. Fertile ova production will be more for large litter size.

ii. Regular heat and conception are more certain.

iii. Gilts should be removed from the breeding pens as soon as they are bred and placed on a restricted diet to reduce embryonic mortality.

iv. The best preparation for breeding consists of exercise / access to lush pasture and reduce grain feeding.

3.17. INFLUENCE OF BOAR CONTACT ON AGE AT PUBERTY IN GILTS

Five minutes of daily physical contact with mature boar is sufficient to stimulate early puberty in gilts. Gilts that are of 165 days of age appear to require daily boar exposure to obtain rapid and maximum pubertal response.

3.18. DETECTION OF HEAT

i. The broad pattern of oestrus is always the same but no two pigs are alike in terms of exhibiting the signs of oestrus. The average length of oestrous cycle in pigs is 21 days (range, 16-25 days).

ii. The oestrous symptoms last for five to seven days beginning with vulval swelling and vaginal discharge.

iii. In true oestrus, there is frequent urination, reduced appetite, mounting and standing for service detected by the erection of ears, and immobility when normal pressure is applied to the back (lumbo-sacral region) or when someone sits on the back and other than this, the application of pressure on the back is also practiced to determine the correct breeding time.

iv. Animals with a predisposition for weak oestrus should be brought near the boar to exhibit heat symptoms a little more clearly and for the effective detection, the herd should be observed in smaller groups.

3.19. TIME TO BREED

i. Eight months of age minimum to breed the gilt

ii. Weight of gilt should be100-120 kg

iii. The length of heat period should be 2-3 days (range 1-5 days)

iv. Gilts: first day heat period and in case of sows: second day of heat are being considered as best time to breed.
v. 2 services for sow at interval of 12-14 hours
vi. Period of oestrous cycle -18-24 days
vii. Oestrus after weaning: 2-10 days
viii. Gestation period:114 days (range 109-120 days)

Since sows ovulate 16 to 48 hours after the onset of oestrus but with the rapid loss of fertility after ovulation. Therefore, it is best to breed during the later half of the first day or early on the second day of oestrus. In many cases the gilts and sows continue to exhibit the 'standing heat' on the next day. In such cases the animals should be re-bred and the interval in the case of re-breeding should be 12 to 14 hours for ensuring high conception rate. The sows should not be bred if they come into oestrus one to four days after farrowing. But can be bred if they come to heat ten days after weaning. The better results can be obtained by breeding them in the second post-lactational oestrus. The animals should be observed for the appearance of subsequent oestrus after breeding. If animals repeat more than twice, it is desirable to cull them from the herd.

3.20. METHOD OF MATING

Hand Mating: Hand mating is the practice of bringing the sows to the boar for individual service when they are on standing heat. In purebred animals, it is the most common method of mating.

Advantages of Hand Mating

i. Use of boar power in best possible way.
ii. Farrowing dates would be accurate in most of the cases.
iii. Repeat breeders can be detected effortlessly.
iv. Limber penis, weak hind legs like breeding defects can be detected easily.
v. Sterile boars are quickly noticed.
vi. Facility for breeding large boars to gilts or small boars to large sows can be made.

vii. Reduced embryonic mortality, if sows are removed from the boar immediately after breeding.

Pasture Mating

i. In this system of mating, the boar is permitted to run with sows and a mature boar can actually serve 15 to 20 sows in a pasture mating system.

ii. One young boar should be used to 8 to 10 gilts in pasture mating system.

Artificial Insemination

i. Advantages: A breeding with additional opportunity of exploiting the effective boar performance.

ii. In a normal ejaculation a boar can ejaculate 200 ml to 500 ml of semen containing 20 to 50 billion spermatozoa. On the availability and provisions of proper storage and semen extenders,15 to 20 sows could be inseminated / or bred per ejaculate.

3.21. CARE OF THE BOAR

i. The boar attains sexual maturity at 6 to 8 months old.

ii. The exotic boar under Indian conditions may take a month or two more to attain sexual maturity.

iii. A 9-month-old boar which exhibits libido should be of use in either hand or pen-mating systems.

iv. Young boars must be used with caution and they should be hand mated (the boar should be taken out of his pen to serve the gilt or sow in an individual pen).

v. A yearling boar can be turned in with 10 to 15 females and can be expected to serve 7 to 8 gilts per week.

vi. The maximum number of services during the breeding season per boar should be four or five per week in case of mature boars and two or three in case of yearling boars.

vii. It is difficult to determine the normal breeding load for boars because of the variation in individual, aggressiveness and ability of the individual boar.

viii. If field- or pen-mating is practiced the following two methods are recommended, viz.

a. To split the sow or gilt herd so as to have dual boar for each group of 15 to 20 sows, and

b. To alternate boars in the sow or gilt herd, i.e. use one boar or a set of boars one day and another‘ boar or set of boars the next day

ix. 30 days of isolation period should be kept for newly inducted boar before breeding. Boars which are not familiar with each other should not be put together in a pen.

x. Outdoor exercise throughout the year is one of the essentials in keeping the boar masculine and in thrifty condition. The boar should be provided with 1·5-2.0m^2 of dry, draft-free, well-ventilated sleeping area.

xi. The tendency to overfeed is a wasteful practice and that results in overweight, decreased libido and sluggishness. The new boar should be fed 2.0 to 2.5 kg of 14 percent protein ration. If the boar is overweight or underweight condition during the breeding season, then the feeding ration should be changed in order to preserve thriftiness and vigour.

xii. Young boars in breeding condition usually grow well on 2-5 kg of feed during the first breeding season and can be maintained with 2 kg when not in breeding. Six to eight weeks prior to the breeding season it should be accustomed by increasing the ration to 2-5 kg.

xiii. A fertility check should be made on each boar at least 30 days before introducing him in the herd for mating. Five to six healthy gilts that are to be marketed and if more than one of these gilts returns to heat within 28 days after mating, the fertility of the boar will be undeniable.

xiv. Hand mating should be done early in the morning and/or late in the evening especially during hot summer season, if the temperature exceeds 27°C. The boar should be fed twice a day, preferably after service.

xv. The two services increase the conception rate by 30 percent on the first heat and the litter size by one and therefore, the sow should be bred at least twice, if possible, during late hours on the first day and early hours on the second day of heat.

xvi. Patience and skills are required in handling the breeding animals, otherwise, the old boars become vicious and young boars become nervous, and they may not mate readily.

3.22. FARROWING MANAGEMENT

i. In a well -managed farms 25 to 30 percent of the pigs born never reach to the market and therefore, the farrowing time is the critical time in pig production.

ii. In the first week after farrowing the death rate is high. Sows may be farrowed in pens equipped with guardrails and a creep space, in farrowing crates or in farrowing stalls.

iii. The pen should be maintained at 24° to 28°C until the piglets are three or four days old and at 18° to 22°C until the pigs are approximately six weeks old. The heat lamps should be hung 45 cm from the floor.

iv. The farrowing pens should be thoroughly cleaned in order to prevent a large number of diseases of piglets.

v. The sows should be washed thoroughly before being brought to the farrowing pen and they should be brought to the farrowing pen at least one week prior' to farrowing so that it becomes familiar with the surroundings.

vi. Then two changes in the ration:

  a. The ration should be made bulky by substituting one-third of the regular ration with wheat bran.

  b. The amount of ration fed should also be reduced by one-third till the sow farrows.

vii. The sow should be watched closely for determining the approximate time of farrowing and feed should not be given 12 hours before farrowing.

3.23. FEEDING SOWS AT FARROWING AND AFTER FARROWING

i. The sows ration should be reduced to nearly half at 4 to 5 days before farrowing in order to prevent constipation and also attempts to be made for inclusion of laxative feeds like wheat bran, linseed meal in the ration.

ii. The amount of high energy feeds like corn to be reduced and to be substituted by wheat bran.

iii. Sufficient amount of clean fresh water should be available throughout the period and it is better to give no feed on the day of farrowing.

iv. It is essential not to feed the brood sow heavily after farrowing for possibilities of producing excessive milk for piglets results in scour. But the sow may be fed the same bulky ration as before farrowing for 3 – 4 days after farrowing.

v. It is best to feed 50% of the ration on the first day and increase the ration by 200 – 300 gms daily in order to reach up to the required amount.

3.24. PREPARATION FOR FARROWING

i. The sow should be isolated from the rest of the herd at about 3 – 4 days prior to farrowing and moderate exercise to be provided while the animals is in farrowing quarters.

ii. Sanitary measures: before moving to the farrowing quarters, the sow should be throughly scrubbed with soap and warm water and the place should be thoroughly cleaned with boiling hot solution or $KMnO_4$ solutions.

iii. Guard Rail – Provision of guard rail around the farrowing pen is an effective means of preventing sows from crushing their pigs. The rail should be raised 8 to 10 inches from the floor and should be 8 – 12 inches from the wall.

iv. Bedding – The farrowing quarters should be bedded with clean, fresh, good absorbent material like with chopped hay and wood shavings.

v. Artificial light – During the time of inclement weather, artificial heat must be provided inorder to maintain the temperature of $55^{o} – 65^{o}$ F during winter farrowing.

3.25. SIGNS OF APPROACHING PARTURITION

Extreme nervousness, uneasiness, enlarged vulva, mucus discharge making a nest for young ones and milk is present in teats are indications in terms of signs of the sow of approaching farrowing.

3.26. FARROWING CARE

i. Many piglets that would otherwise be lost can be saved if someone is on hand to assist and therefore an attendant should be on hand.
ii. It takes generally two to four hours for complete farrowing to take place and placenta drops one after the other within two hours after completion of the farrowing process.
iii. The piglets should be removed as they are farrowed and kept warm in the creep space until farrowing is complete.
iv. Each piglet should be cleaned of all mucus to ensure that the breathing passages are clear.
v. The navel cord should be tied 2.5 cm away from the navel cut with a disinfected pair of scissors, and the stumps painted with iodine. This section will soon dry up and drop off leaving a clean non-infected navel.
vi. Piglets should be allowed to be nursed after birth. In about two days they settle down to their individual teats.
vii. They nurse eight to ten times in 24 hours in the initial period.
viii. Trampling by the sow should be prevented during the first two weeks.

3.27. FARROWING DIFFICULTIES

i. The major troubles at farrowing are mastitis and that must be treated with effective antibiotic immediately.

ii. Agalactia: the causes of agalactia are not always same but may be due to insufficient secretion of oxytocin, that results in poor secretion of colostrum. This may be treated by the intravenous injection of 5 to 10 IU of oxytocin. Failure of milk secretion may be due to constipation, which may be cured by administration of suitable laxative like caster oil.

iii. The inflammation or injury of vulva at the time of farrowing may cause fever, which may also cause reduced feed intake and subsequently results in poor secretion of milk.

iv. Under such conditions like agalactia, mastitis and inflammatory diseases, it is better to rear the piglets by foster mother or by hand feeding.

3.28. MANAGEMENT OF DIFFERENT CATEGORIES OF PIGS

Care and Management of Suckling Sows

i. Sows should be fed more during suckling period rather than the pregnancy period since the brood sow needs feed not only for its body maintenance but also for the development of mammary gland in order to provide optimum quantity of milk to nurse the piglets of large litter size. Th sow generally gives about 2.5 to 4 kg of milk per day.

ii. A good sow should give about 150 to 200 kg of milk during the 8-week suckling period. Sow's milk is much more concentrated than cow's milk.

iii. The sow is greedy by nature in terms of feeding and therefore, it should be fed to avoid over weight and also care must be taken to maintain the normal healthy weight.

iv. A sow requires about 4 to 6 kg of feed daily depending on its nursing ability and litter size.

v. A practical hint to be followed is to allow 1.5 kg of feed for the sow and add 0.5 kg of feed per every piglet to a maximum of 5 to 6 kg of the total ration.

vi. Ten days after farrowing, the sow may be allowed to graze along with its litter. Good quality as well as tender pasture is most desirable for suckling sows.

vii. Sufficient lucerne hay or succulent fodder may be provided to the sow, if it is not allowed on pasture.

viii. The concentrate feed meant for piglets should be protected by creep arrangement to prevent the sow from eating it.

ix. The feed should contain 14 percent protein for mature sows and 15 percent protein for gilts.

x. Maize or other grains, groundnut cake, fishmeal, meat scrap, dairy products etc., may be included in the ration.

xi. The feed should be sufficient in various minerals and vitamins to manage any deficiency in the sow or the piglets.

xii. A few days prior to weaning, the quantity of feed should be gradually reduced so as to restrict the milk flow in order to dry out the udder but further complications like mastitis should be avoided.

Care and Management of Piglets

i. The baby pig begins to explore his environment within a few minutes after birth and soon finds his way to nipple and begins to suckle.

ii. It is essential that the pen environment should be clean to minimize chance of exposure to disease and parasites.

iii. Normally individual piglets identify themselves with a particular teat during the first few days of life. The teat order will be set within 10 days. But in small litter it is common for a piglet to claim more than one nipple. On the other hand, in larger litters, the weakest pig may starve because they cannot compete for a nipple.

iv. The normal nursing behavior pattern is for a uniform time interval between each feeding over a 24 hours period. The average nursing interval is less than 1 hour so that suckling piglets receives more than 24 feeding daily, each feeding consist of only few minutes and it decreases when lactation advances.

v. To prevent 'naval ill' the naval card should be tied off to prevent loss of blood and it should be cut 3-5 cm distal to the legation and this portion should be dipped in a solution of 2% iodine or 70% ethyl alcohol.

3.29. RAISING ORPHAN PIGLETS

i. The death of a sow after farrowing is mainly due to lactational failure and mastitis.

ii. The orphan piglets are raised by the use of foster mother or foster sow or by use of milk replacers.

iii. In the foster sow technique, transfer of orphan piglets must be made within few days after farrowing.

iv. To ensure the acceptance of new pigs, the sow should be separated from her own litter for a short time and the new piglets are brought to the sow.

v. A good sow milk replacer consists of one egg yolk thoroughly mixed with one litre of cow's milk. This mixture supplies well balanced diet.

3.30. REMOVAL OF NEEDLE TEETH

i. Piglets are born with four pairs of sharp teeth with two pairs on each jaw and are known as wolf teeth.

ii. This tooth has no practical value to piglets and they irritate the sow's udder during nursing or cause injury to other piglets at the time of nursing. So, it is a good practice to clip the needle teeth shortly after birth (0 – 3 days).

3.31. EAR NOTCHING OF PIGS

i. Pigs should be ear notched at the time of birth.

ii. Ear notching is required for registering pure bred pigs for identification.

iii. The pig ear is considered in two halves.

iv. Four areas are available for notches.

v. Each of the notches is assigned a number, thus notch in the lower edge indicates the number 1, while a similar notch in the lower edge of the tip half indicates number 3.

vi. Each notch in the upper edge of the tip half count as number 9 and notches on the base half are 27; a notch in the tip of the ear has a value of 81.

vii. Combination of these notch numbers gives a wide variety of numbers for litter identification. The right ear is used for litter mark and left ear is used for pig number.

3.32. PIGLET ANAEMIA

i. Prevention of anaemia is necessary for all the pigs kept in confined system of rearing. If the piglets are reared on pasture, they will get iron from the soil. Pigs reared in confinement should be treated for the piglet anaemia at 0 – 3 days of age with an injection of 100 – 150 mg of iron dextran injection at the side of the neck.

ii. The piglet anaemia condition can also be prevented by oral administration which consist of spraying, swabbing in sow's udder with saturated solution of ferrous sulphate (0.5 kg of ferrous sulphate in 10 litters of hot water) so that the piglet will get some iron in each time during suckling milk.

3.33. CASTRATION OF PIGLETS

i. The male piglets may be castrated early, if they are not kept for breeding puposes.

ii. Castration may be performed at anytime but the correct time to perform castration at 3 – 4 weeks of age as because the controlling the animal is relatively easy.

iii. Pigs are castrated to maintain the quality of meat and also to prevent the boar odour in the cooked meat of uncastrated males. By this way the uncontrolled breeding can also be prevented.

3.34. CREEP FEEDING

i. Piglets begin to develop an appetite for the dry feed at the age of 2 – 3 weeks.

ii. Provision of additional nutrients at this time is essential for the maximum growth and other physiological and reproductive developments are to be achieved during the period from farrowing to weaning.

iii. The creep feed should be palatable, rich in nutrients and should be easily digestible.

iv. The creep area may be provided to all piglets in the farrowing pen.

3.35. WEANING, IDENTIFICATION AND DEWORMING OF PIGS

Weaning

i. Normally piglets are weaned at 8 weeks of age.

ii. Weaning at an early age is a specialized job and should be practiced only by experienced person.

iii. The standard weight of weaning of piglets is 6 kg and above. Normally in pig farm piglets, are weaned by weight and not by date.

Identification

Weaning is the ideal time for giving identification to the piglets by tatooing and tagging. The common method of identification of piglets is ear notching at the age of 10-15 days old. The tattooing is done by using tattooing forceps at the inner side of the ear devoid of blood vessels. Tagging is being performed by plastic tags, brass tags and polyurethane tags.

Deworming

i. The piglets should be dewormed at the time of weaning or at 1 or 2 weeks after weaning.

ii. The ideal drug to deworm the piglets is albendazole given at the dose rate of 1 ml / 5 kg body weight.

3.36. FEEDLOT STOCK MANAGEMENT

Fatteners

i. Very few piglets are raised for bacon or for lard purpose in tropical and mostly they are reared for pork production and are being marketed after attaining 45-63 kg of body weight.

ii. The fattening stock should have uniform size, age and it is better to fatten litter mate.

iii. Normally the fatteners attain 54 kg by 5th month of age. They should again be drenched with piperazine by 15th to 16th weak to control internal parasites.

iv. The feed should be in the form of dry or wet. The feeding should be at regular interval otherwise the pigs may be restless and may loose some energy

v. They should be fed to their appetite but not to repletion

3.37. WEANING OF PIGS AND GROWER MANAGEMENT

i. Normally piglets are weaned at the age of eight weeks [56th day]. Weaning at an early stage is a specific job and should be practiced only by experienced persons.

ii. At weaning the sow should be taken away from the piglets and not the piglets from the sow.

iii. Weaning should be a gradual and not an abrupt process. At first the sow should be taken away for few hours, then for whole day and finally for all time.

iv. The grower piglets should be grouped according to the size. The group size may vary between 10 and 20.

v. Optimum floor space should be provided to minimize tail biting.

vi. The piglets should be shifted from 18% protein creep ration to 16% protein grower ration.

vii. The piglets should be given broad spectrum anthelmatic drug such as piperazine immediately after weaning in order to control internal parasite.

3.38. BASIC REQUIREMENTS

i. Location of the pig sty, floor space requirement, feeding and watering are to be taken in to consideration.

ii. Construction details of pig farm

iii. Other facilities required like farrowing pen, guard rails, wallowing tank etc.

iv. Farrowing two litters per sow annually should be attempted. The annual cost of maintaining a sow varies little whether it farrows one or two litters per year. Adequate housing and equipment for raising pigs are necessary to provide shelter against inclement weather, prevent diseases, control parasites and save labour costs.

v. The pattern of houses suitable for a particular area may not be suitable for other areas. However, certain basic considerations in providing housing for the herd should be designed in a way to give maximum comfort to pigs for their better growth. Dampness, draft and overheating should not be allowed.

vi. Every effort should be made to utilize the locally available materials to make it cost effective. The flooring should have a rough finish and with regular masonary type made up of waterproof cement mortar or lime mortar. Proper drains should be provided so that the effluents are disposed off.

vii. Under village conditions, the housing can be made up of pens measuring 3m x 2.4m or 3m x 3m, with an open yard of nearly the same dimension or in some cases slightly longer. Walls should be constructed 1.2 to 1.5 m high from the floor.

viii. Some of the pens could be converted into farrowing pens by providing guardrails made up of G.I. pipes of 5 cm diameter, along the walls, 20 to 25 cm from the ground and wall. In addition to guardrails, creep space can be provided for the piglets along the wall by making a partition or in one of the corners with separate entrances for the piglets. This space is usually of 0.75m x 2.4m area.

ix. the yard with regular flooring to be avoided and it should be the open yard portion of the housing as a natural ground connected to slope. In few farms, the regular flooring is often made on grounds of sanitation for the control of diseases, particularly in places where exotic pigs from temperate regions are reared.

x. As the piggery-development programme envisages the use of exotic breeds, their management under conditions prevailing in this country like India needs special attention.

xi. Prolonged exposure of the animals to bright sunshine may cause swine to become overheated, even during moderate weather. Shade helps in preventing deaths and increasing production efficiency during hot weather. It is desirable to plant trees in the neighbourhood of pens for reducing the intensity of heat, it is not desirable to plant trees for giving regular shade, because they permit rapid build-up of parasite and dampness.

3.39. LOCATION OF THE PIG STY

i. The pig sty should be located on elevated and well drained site not liable to flooding. The structures should be at least 15 m away from the human dwellings and factories, 30 m away from the dairy farm, poultry houses and food grain storage.

ii. 40 m away from the fire sources, 1 km away from the garbage dumping grounds, slaughter house and leather factories. The site should have adequate water supply, proper sunlight and air circulation.

3.40. TERMINOLOGY OF PIG STY

Sty: A dwelling place consists of one or more pen for pigs.

Pen: It is a part of pig sty for housing single or group of pigs having covered place and an adjoining open yard. The covered place providing feeding and watering trough and also dunging space.

Dry Sow and Gilt Sty: It is a dwelling place for dry sow and gilt.

Farrowing Sty: It is a dwelling place for nursing sows and litter.

Fattening Sty: It is a dwelling place for young pigs kept for fattening purpose.

Sick Animal Sty: It is a dwelling place for sick pigs particularly meant for segregated from healthy stock.

Stud Boar Sty: It is a dwelling place for breeding boars.

3.41. SPACE REQUIREMENTS

Covered floor area requirement:

Sl.No.	Type of animal	Covered area per animal (m^2)
1	Boar	6.25 – 7.5
2	Farrowing sow	7.5 – 9.0
3	Weaner Fattening pigs	0.96 – 1.8
4	Dry sow gilt	1.8 – 2.7

Open yard requirement:

Sl.No.	Type of animal	Covered area per animal (m^2)
1	Boar	8.8 – 12.0
2	Farrowing sow	8.8 – 12.0
3	Weaner Fattening pigs	8.8 – 12.0
4	Dry sow Gilt	1.4 – 1.8

Feeding and watering trough area requirement:

Sl. No.	Type of animal	Feeding space/ animal
1	Boar	60 – 75 cm
2	Sow with litter	75 – 100 cm
3	Fattening pigs	30 – 40 cm
4	Weaner	20 – 30 cm

In all the types, the width of the trough either feed or water trough shall not be less than 50 cm and the watering space must be 1/10 of feeding space.

3.42. FARROWING STY AND GUARD RAILS

Farrowing Sty:

i. The farrowing sty shall be designed to provide sufficient room for nursing sow to sleep and for the piglets to move around. It should be provided with one or two rows along the walls of the farrowing sty compartments for the piglets.

ii. Each compartment should measure 2.5 x 0.75 x 1.2 m height and opening of 300 x 225 mm space shall be provided in each compartment facing the nursing sow.

iii. Watering and feeding troughs need to be provided in each compartment. Guard rails shall be provided 25 cm above the floor level.

iv. Guard Rails: It is made up of GI pipe of 5 cm diameter and it can be fixed along the wall 20 – 25 cm from the ground and wall.

3.43. CONSTRUCTION DETAILS OF PIG STY

i. Floor: The floor of the pig sty should be hard and impervious to water and easy to clean and the floor should be laid with concrete and slope towards the drainage.

ii. Roof: The roof may be RCC flat type or gable roof. Materials for roofing may be asbestos cement sheets or corrugated steel sheets. In the regions of extreme climatic conditions, the roof may be insulated by providing a layer of thatching to reduce extreme heat.

iii. Door: The number of doors provided in each sty shall depends on the dimensions of the house. Door made up of iron angle and the width of the individual door to individual pen may range from 0.75 – 1.00 m.

iv. Drainage: Each pig sty shall be provided with alleast one drain on each pig pen. The drain shall have a minimum width of 250 mm and a slope of 25 mm.

3.44. WALLOWING TANK

i. Pigs have got few sweat glands and in areas having warm weather, the mature breeding animals and fattening animals need a wallow during summer months. Instead of permitting unsanitary wallows, a masonary wallow with proper drainage system is desirable.

ii. The size of the wallow will depend upon the number and size of the animals. Wallowing tank may be provided in pig breeding farm suited to the climatic zones.

iii. The wallowing tank made of cement concrete and dimension of the tank is less than 2.5 m x 1.2 m x 0.15 m. alternatively overhead sprinklers and showers may be provided in order to keep the animals cool during summer.

3.45. FUNCTIONAL UNITS OF STY

i. Boars sty
ii. Dry sow and gilt sty
iii. Farrowing sty
iv. Weaner / grower sty
v. Fattening sty.

BOAR STY: Each boar sty shall normally, have not more than 24 pens under one roof and each pen shall accommodate not more than one animal. 6–7-meter square per boar under pen and 9-meter square per boar in the run area

DRY SOW AND GILT STY: It is meant for growing gilt or dry sow. Each sty shall normally have not more than 40 pens under one roof and these pens shall be arranged in two rows. 5-10 pigs can be accommodated in each pen depending upon the stage of pregnancy and age.

FARROWING STY:

i. Pregnant sows are transferred to these sties 7-10 days before the expected date of farrowing and left here up to weaning of their litters (56 days).
ii. Each farrowing sty shall normally contain not more than 40 pens under one roof and each pen shall accommodate not more than one animal with the litter.
iii. Number of farrowing pens required is 10 % of the breedable female.
iv. Floor sapce allowance is 9 square meters both in pen and run
v. Each pen should have the facilities of guard rail and creep area

Guard Rail: The guard rails are made up of tubular iron rod of 2 to 2.5 inches thickness and fixed 25 to 30cm away from the wall and above the floor. So, that the piglets can run under the rail when the sow lies down so that the crushing of piglets can be avoided in between the sow and walls.

Creep Area: In corner of the pen creep area should be provided with the help of stone slab should be fitted in the corner of the pen, 25-30cm above the floor level. In that area piglet only can enter and get the creep feed. Lighting can also be provided in creep area to give warmth to young ones. Since piglets having thin layer of subcutaneous fat, it is more susceptible to cold stress.

Bedding: For good results 4 to 6" bedding is needed especially in winter and at first 3 days after farrowing to give warmth, softness and non -slippery floor. It must be kept clean dry and evenly distributed. Wallowing tank should not be provided in the farrowing pen in order to avoid trampling of piglets in to the wallowing tank.

Weaner Sty

i. Piglets that are weaned are housed in these pens up to six months of age.

ii. Each sty shall normally, have not more than 30 pens under one roof. Each pen may accommodate 30 piglets of age group between two to six months.

iii. Number of weaner/grower sty is equal to the number for breedable female in the farm.

iv. 0.9 to 1.8 square meter per animal under pen and run.

Fattening Sty

i. These are for housing pigs that are being transferred for the market.

ii. Each fattening sty shall normally have not more than 20 pens under one roof.

iii. Each pen may have 16 to 32 animals kept for fattening purpose.

3.46. STY FOR SICK ANIMAL

The number of pens under this sty shall depend upon the number of animals kept in farm normally a minimum accommodation for 5 percent of the stock shall be provided for housing the sick animals.

3.47. SUBSIDIARY ITEMS/ACCESSORY ITEMS

Apart from the building for the stock, a few subsidiary buildings like weighing yard, store buildings, feed store, loading and unloading ramp, manure pit, garbage boiling vat and boiler room and wallowing tank.

Weighing Yard: The specification for the weighing yard may vary from one place to the other. A weighing shed shall be provided where weighing scale shall be fitted to record the weight of the animals regularly.

Store Building: The store buildings may have a store-keepers office, a mixing room to mixture feed of the animal, a room for keeping farm equipment tools.

Feed Store: The size of the feed room shall vary according to the number of animals kept and it is used to storage of feed items.

Manure Pit: The manure pits shall be constructed at a distance of not less than 15m away from the sties. As regards the liquid manure, including urine and washings a suitable provision shall be made for its disposal for irrigating the cultivated areas.

3.48. FORMULATING PIGGERY PROJECTS OF VARIOUS SIZES

The entrepreneurs and farmers wish to know the economic viability and reliability of the pig farming enterprise before entering into the business. Moreover, the availability of finance from institutional sources for commercial pig farming will be crucial for its development. Again, to avail such finance, it is essential to develop and submit a technically feasible and economically viable project proposal on pig farming. Hence here is an attempt to generate information on techno-economic feasibility of commercial pig farming project by taking an example of 10-sow unit. The schemes for smaller or larger farms may be formulated taking this example as a model project.

General Assumptions

i) Constant breeding stock of 10 breedable sows is maintained. Breeding is approximately 100%. The parent stock may be replaced by purchased animals after 8-9 farrowings.

ii) Death in breeder stock will be compensated by insurance claim.

iii) As the project is assumed for 5 years, 9 crops of piglets will be sold out in 60 months.

iv) Piglets are weaned from mother after 2 months and half of each litter is reared upto market age, i.e., 8 months and sold out. The other half of the litter is sold out at weaning.

v) The incremental rate in price in input would be nullified by that in price of meat/live animals.

vi) Foreseeable planning horizon for bankability purposes would be 5 years.

vii) The predominant share of income would be accruable from only domestic meat market.

Technical Norms

Sr. No.	Particulars	
1	No. of sows (6-7 months old)	10
2	No. of boars	1
3	No. of batches	2
4	Interval between two batches (months)	3
5	No. of farrowings per year	2
6	No. of piglets per sow per farrowing	11
7	Mortality among piglets (weaners)	20%
8	Mortality among fatteners	10%
9	Mortality in adults is not considered as insurance cover is available	
10	Weaning period (months)	2
11	Space requirement (s.ft.)	
	a. Boar	70
	b. Lactating sow with it's piglets	100
	c. Dry sow	20
	d. Fattener of 3-5 months age	10
	e. Fattener of 6-8 months age	15

Economic Norms

Sr. No.	Particulars	
1	Total area required for building (Sq. ft.)	1090
2	Store room (s.ft.)	100
3	Concentrate feed requirement (kg/day)	
	a. Breeding stock	2.5
	b. Weaner	0.3
	c. Fattener (3-5 months age)	1.5
4	Total area required for building (Sq. ft.)	1090
5	Green fodder (kg/day/adult)	3
6	Cost of construction of sheds (Rs./s.ft.)	300
7	Cost of construction of store room (Rs./s.ft.)	500
8	Cost of boar (Rs.)	6000
9	Cost of sow (Rs.)	5000
10	Cost of concentrate feed (Rs./kg)	10
11	Cost of green fodder (Rs./kg)	1.5
12	Insurance (%)	6
13	Cost of medicines and vaccines (Rs./pig)	100
14	Misc Cost {power, water, other misc. expenses (Rs/pig)}	150
15	No. of labourers required	1
16	Labourer wages (Rs. Per month)	3000
17	No of piglets sold per sow per farrowing (2 months old)	4
18	No. of fatteners sold per sow per farrowing (8 months old)	4
19	Sale price of piglet (Rs./piglet)	2000
20	Avg. wt. of fattener (kg.)	80
21	Sale price of fattener /parent stock (Rs./kg)	70
22	Closing stock value of parents	150% of original value
23	Income from manure	
	Weaner/fattener (Rs./month)	20
	Adults (Rs./month)	50
24	No. of gunny bags per ton of feed	20

[Table Contd.

Contd. Table]

Sr. No.	Particulars	
25	Sale of gunny bags (Rs./bag)	10
26	Depreciation on sheds (%)	5
27	Margin money (%)	15
28	Interest rate (%)	12
29	Repayment period (years)	5
30	Grace period (years)	1

Economic Analysis of the Project

Details of investment		
a.	**Capital cost**	
	Pig sheds and other structures	**Rs. 3, 77,000**
	Farrowing pens (4) for lactating sow	Rs. 120000
	Boar cum service pen	Rs. 21000
	Dry sow pens (6)	Rs. 36000
	Fattener shed –I (20)	Rs. 60000
	Fattener shed –II (20)	Rs. 90000
	Store room	Rs. 50000
	Water supply system	**Rs. 20,000**
	Breeding stock: Crossbred gilts 10 Nos. @ Rs. 5000 each and Crossbred boar 1 No. @ Rs. 6000 each	**Rs. 56,000**
	Total capital cost	**Rs. 4, 53,000**
b.	**Recurring cost**	
	Capitalized recurring expenses upto 2 months	Rs. 2, 01,940
c.	**Total project cost**	
	Capital cost	Rs. 4, 53,000
	Recurring cost capitalized	Rs. 2, 01,940
	Total capital cost	Rs. 6, 54,940
	i) Margin money (15% of total capital cost)	Rs. 98,241
	ii) Net finance required from bank	Rs. 5, 56,700

Sr. No	Particulars	Year I	Year II	Year III	Year IV	Year V
A	**Fixed cost**	453000				
B	**Reccuring Cost**					
1	a. Breeder Feed					
	Concentrates	88500	100375	100375	100375	100375
	Fodder	15930	18068	18068	18068	18068
	b. Weaner feed	14400	28800	28800	28800	28800
	c. 1st batch	27000	144000	81000	81000	81000
	d. 2nd batch	-	81000	144000	144000	144000
2	Insurance (6%)	3360	3360	3360	3360	3360
3	Labour cost	36000	36000	36000	36000	36000
4	Medicine cost	12100	23100	23100	23100	23100
5	Misc expenses	4650	19650	19650	19650	19650
	Total	**201940**	**454353**	**454353**	**454353**	**454353**

Income

Sr. No	Particulars	Year I	Year II	Year III	Year IV	Year V
1	Sale of piglets @ Rs. 2000	80000	160000	160000	160000	160000
2	Sale of fatteners (70x80 kg)	-	448000	448000	448000	448000
3	Sale of gunny bags	2600	7080	7080	7080	7080
4	Sale of manure	7050	16200	16200	16200	16200
		89650	631280	631280	631280	631280
	Total	**89650**	**631280**	**631280**	**631280**	**631280**

Cash Flow Analysis

Sr. No	Cost	Year I	Year II	Year III	Year IV	Year V
1	Capital cost	453000				
2	Reccuring cost	201940	454353	454353	454353	454353

[Table Contd.

Contd. Table]

Sr. No	Cost	Year I	Year II	Year III	Year IV	Year V
3	Depreciated value of shed	0	0	0	0	299730
4	Closing stock value of animals	0	0	0	0	84000
5	Total - 89650	631280	631280	631280	1015010	
6	Profit - 112290	176927	176927	176927	560657	

Repayment Schedule

	Year I	Year II	Year III	Year IV	Year V
Income	89650	631280	631280	631280	1015010
Expenditure	*	454353	454353	454353	454353
Gross surplus	89650	176927	176927	176927	560657
Loan balance	556700	417525	278350	139175	0
Equated annual installment	0	139175	139175	139175	139175
Interest	66804	50103	33402	16701	0
Net surplus	22846	-12351	4350	21051	421482

*as recurring cost is included in the loan amount.

Total net income in 5 years : Rs. 4, 57,378

Net income per annum : Rs. 91,476

Net income per month : Rs. 7,623

Economic Evaluation of the Project: Economic evaluation of the present project is carried out using Net Present value (NPV), internal rate of return (IRR) and Benefit cost (B/C) ratio. The prevailing bank rate (@12%) is taken as the discounting factor to calculate the present value and IRR.

Net Present Value (NPV) : Rs 53292.45

Internal rate of return (IRR) : 166%

Benefit cost ratio (B/C) : 1.21:1

Economic/Financial Viability: Total initial fixed cost for the project of 10 sows will be Rs. 4, 53,000. If the recurring cost for the first 12 months is considered as capitalized cost, then the total capital requirement for the project would be of Rs. 6, 54,940. Considering 15% of the total cost as margin money, Rs. 5, 56,700 will be required as net bank loan. The project will generate net returns of Rs. 4, 57,378 over the period of 5 years planning horizon, giving net income of Rs. 91,476 per annum and Rs. 7,623 per month. NPV of the project is Rs 53292.45 which is quite high and IRR as 166% also is very high. The above two indicators imply the economic and financial soundness of the project. Therefore, the above stated scheme is technically feasible and economically viable.

CHAPTER 4

NUTRIENT REQUIREMENTS AND FEEDING OF PIGS

Dr. Bikas Chandra Debnath and Dr. Tapan Kumar Das

4.1. INTRODUCTION

The pig industry in India has not yet been developed like poultry, dairy and wool industries. It is mainly due to the fact that pig is not used as meat by majority of meat eaters. It is estimated that the production of pork constitutes only 5% of the total meat produced in India. But the consumption of pork in India has greatly increased in the recent years. The grading up programme of non-descript pig with exotic breed of Large White Yorkshire; Landrace etc. is dramatically changing the pig population. Nowadays a number of piggeries have come up where hogs are reared on scientific lines. So, it is necessary to know about the nutrient requirements and principles of feeding this important meat producing animal especially when feed represents about 70-75% of the cost of the inputs in the hog industry.

4.2. SOME IMPORTANT FACTS ABOUT PIG

- Highly prolific among meat producing animals.
- Gestation period is 114 days
- 5 times its birth weight at the end of 3 weeks and 10 times its birth weight at the end of 5 weeks may be achieved.

DOI: 10.1201/9781003749844-4

4.3. NUTRIENT REQUIREMENTS FOR PIGS

Because of group feeding nature of the pigs, nutrient requirements for them are expressed as percent (%) of the diet or amount per kg ration. The latest BIS specifications (1986, reaffirmed in 2001) are on 100% DMB where as NRC, USA specifications are on 90% DMB. The latest NRC (2012) has given the nutrient requirements for 7 various weight groups of growing finishing piglets viz. 5-7kg, 7-11kg, 11-25kg, 25-50kg, 50-75kg, 75-100kg and 100-135 kg. The amino acid requirements for the pigs can be expressed in a better way by grouping such body weights. In our country, for easy understanding of feeding of pigs, they are grouped into three body weight groups viz. 2-12 kg (up to 56 days/ 2 months); 12-50 kg body weight i.e. growing period (2-4 months) and 50-70 kg i.e. finishing stage (4-6/7 months). In India, pigs are reared up to 70 kg body weight before it is marketed for meat purpose whereas in foreign countries, they are reared up to 90 kg body weight before it is marketed for same purpose.

Various ratios according to growth stage of Pig (BIS)

Body weight (kg)	Time required	Stage of growth	Name of ration
2-12 kg	Up to 56 days.	Starting	Starter
12-50 kg	2 months to 4 months	Growing	Grower
50-70 kg/90kg	4 months to 6/7 months	Finishing	Finisher

4.4. NUTRIENTS FOR PIG FEEDING

Carbohydrates

Pig diet is mainly composed of carbohydrates which is around 75% of the total dry matter. Pigs can handle sugars and starches very efficiently. However, they can utilize crude fibre to a lower extent. The capacity of digestive tract of pig is small as compared to the ruminants. As they are monogastric animal they do not have cellulytic enzyme to digest the cellulose. However, caecum and colon of pigs have microbial population having cellulose activity due to which a small amount of cellulose is digested in the large intestine. The end product of cellulose digestion and fermentation is volatile fatty acids which are absorbed directly from caecum and colon.

The utilization of fibre by the pigs depends on the age and weight of pigs and characteristic of non-fibrous portion of the ration. The stage at which the fodder is harvested for feeding to the pigs also makes a difference. The crude fibre of plants harvested at an early stage will have better digestibility than harvested at later stage. Heavier the pig the more fibre (cellulose) it can use. For growing and finishing pigs 5-6% crude fibre level in their diet is recommended. In sows, 10-12% level of CF in diet can be well tolerated.

4.5. ENERGY REQUIREMENTS FOR PIGS

The amount of feed consumed daily by the growing finishing pigs, allowed feed *ad libitum*, is controlled principally by the energy content of the diet, provided it is balanced with all essential nutrients. The rations rich in energy are consumed less per unit of body weight than the rations poor in energy. By controlling the energy in take we can alter the fat: lean meat ratio. The present day's trend is towards meat type hogs and for this restriction of energy is must.

Like poultry, energy requirements for pigs are also expressed as Metabolizable energy (ME) per kg diet. ME is in fact 94-97% of DE (average 96%). The energy lost in gas production in the digestive tract is only 0.5-1% of DE. The relationship between DE and ME depends upon both the quality and quantity of protein. A poor quality protein leads to decreased ME value. If excess protein is supplied to the animal then also ME decreases because excess amino acids are not used for protein synthesis are catabolized and used as a source of energy and the N is excreted as urea. So, N content of urine increases and ME value decreases.

Energy Requirements During Pregnancy

If the pregnant sows are allowed to feed *ad libitum* then they will consume more energy during gestation than they require for maintenance and for development of foetus. Therefore, energy restriction is must. The energy restriction during gestation has little effect on either litter size or birth weight. DE requirement for maintenance (DEm) for gestating sow is 100 Kcal/ kg metabolic body weight/ day. Sows should gain a net of 25 kg throughout pregnancy for 6-10 piglets. The weight of product of conception and placenta should be approx. 20 kg, for total of 45 kg of gestation weight gain of sow. Feed intake of 1.7-2.3 kg per day during pregnancy has no significant effect on the total number of piglets born.

Energy Requirement During Lactation

Sow milk is very rich in energy and it contains: 8.3% fat, 5.5% protein, 5% lactose, 11.4% SNF, 0.7% ash. The Ca and P content of sow milk are 2.5g and 1.7g/kg of milk, respectively. The net energy requirement of the lactating sow for dietary energy is the sum of that expended in maintenance plus the gross energy of milk less the contribution made by mobilized body tissue. Hence, lactating sows are fed at a level to support maintenance plus requirements for milk production. Energy requirements for lactating sows are to be adjusted in relation to the litter size.

Maintenance requirements of sows (ME, Kcal/day) will be 105 Kcal ME/kg metabolic body weight, with the requirement of boars being 15 percent higher. It includes energy requirement for maintenance and lactation. Maintenance requirements of growing-finishing pigs (ME, Kcal/ day) will be 197 Kcal ME/kg metabolic body weight. DE requirement for maintenance (DEm) for lactating sow is 110 Kcal/ kg metabolic body weight/ day. DE requirement for maintenance (DEm) for gestating sow is 100 Kcal/ kg metabolic body weight/ day. Hence, the energy requirement of a lactating sow is 5-10% more than that of the gestating sow.

Energy requirement for milk production is 2000 Kcal of DE/ kg milk. It can be calculated by assuming that gross energy of milk (NE) is 1300 Kcal/ kg of milk. That means efficiency of utilization is digestible energy for milk production is 65%. If the contribution of milk is from the body weight loss then 30% weight loss is due to fat and 13% weight loss is due to protein. So, the weight loss will contain an energy content of 4500 Kcal/ kg which is converted in to milk with 85% efficiency (NRC, 1998).

4.6. PROTEIN REQUIREMENTS

In pig feeding, it is important to provide good quality protein in the ration. All essential amino acids should be present in right quantity and proper balance. Even if one essential amino acid is lacking or is in excess it will cause marked reduction in the feed intake which will affect the growth and production. A combination of animal and vegetable protein in a pig's diet will provide all essential amino acids in proper proportion.

Protein requirements of pigs in general in the feed is given below:

Class of Pig	% of protein in diet
Pigs – pre-weaning/creep feed	18-22
Weaned pigs	16
Growing pigs at 45 kg body wt	14
Breed gilts	15
Sows	14
Breeding boars	14
Lactating sow	15

For pig feeding, both quantity and quality of protein is very important. For the optimum performance the essential amino acids must be fed at proper level, ratio and time. The vegetable proteins from grains and the by-products of concentrates would provide the major portion of the protein as well as amino acids requirements. But these proteins are likely to be deficient in lysine, methionine + cystine and tryptophan. Animal protein sources are rich in lysine, methionine and tryptophan. So, a pig diet should be supplemented either with natural proteins rich in these amino acids like fish meal, meat meal, blood meal, etc. or with the synthetic amino acids. which are available in market. Fish meal should not be fed at levels higher than 10% as otherwise objectionable fishy flavour may develop in the pork.

Protein Requirement for Maintenance

A value of 8.97g protein/kg metabolic body weight/day is required for the maintenance of pig. An apparent digestibility of protein is 80%. Overall efficiency of protein for body gain is 70% of digestible protein.

Protein Requirement During Lactation

An apparent digestibility of protein for pig is 80%. Overall efficiency of protein for milk production is 50% of digestible protein. Estimates of gross energy efficiency of conversion of dietary protein to milk protein vary from 0.30 to 0.45. The net protein requirement for a lactating sow is made up of that for maintenance plus that secreted in milk. Gross efficiency of protein utilization is 0.40. A sow producing 10 kg milk/day would secrete (10×55) = 550g of protein per day in her milk. Now, the CP requirement will be 570/0.4=1375 g.

4.7. AMINO ACID REQUIREMENTS

Pigs can synthesize arginine at a rate sufficient to meet about two-third of the requirement of normal growth. So, small amount must be provided in the diet. Cystine can satisfy 50-70% of the total need for sulphur containing amino acids (methionine + cystine) and methionine can meet the total need in the absence of cystine. Phenylalanine can meet the total requirement of phenylalanine + tyrosine as it can be converted to tyrosine but tyrosine cannot be converted to phenylalanine it can satisfy only 30% of the total requirement of the two amino acids.

Clinical Signs of Amino Acid Deficiencies or Imbalances in Pig

Deficiency causes a reduction in feed intake with increased feed wastage, impaired growth and general unthriftness. Excess protein is detrimental to the animals which may lead to scouring and enlargement of kidneys, liver and spleen.

4.8. FAT AND ESSENTIAL FATTY ACIDS REQUIREMENT FOR PIGS

When high fat diets are fed to pigs there is deposition of excessive fat inside the body. If this feed fat contains higher concentration of short chain fatty acids (ex. Soybean and groundnut) there is a production of soft pork which is not desirable. A pig fed on ration containing 0.5% fat, makes a satisfactory gain and stores normal amount of body fat. However, the practical level of fat inclusion is higher than this i.e. about 4%. A diet containing 0.6% fat has produced signs of the fat deficiency symptoms in weaning pigs: loss of hair, scaly dandruff like dermatitis, necrotic areas on the skin around neck and shoulders, unthrifty appearance, retarded sexual maturity, under-developed digestive system, poor development of the gall bladder, slow growth rate, lower efficiency of feed conversion and enlargement of thyroid gland. These symptoms appear after 63 days of age (9weeks). Pig requires fatty acids for their growth. Among the important essential fatty acids, lenoleic acid and arachidonic acids are important. Lenoleic acid is dietary essential as it is not synthesized at all by the animal tissue.

4.9. REQUIREMENT OF MINERALS

Calcium and Phosphorus

If pig is fed on concentrates alone calcium is more likely to be deficient. Whereas if they are fed only on pasture, phosphorus deficiency results. The recommended calcium and phosphorus levels for pig diet are given in table:

Calcium and phosphorus requirement of pigs according to Braude (1978)

Category of pig	Body Weight	Ca % on DMB	P % on DMB
Growing & Fattening Pigs	(up to 20 kg kg)	1.14	0.91
	(20-55kg)	0.90	0.69
	(55-90 kg)	0.80	0.63
Gilts & Sow bred		0.75	0.50
Lactating Sows and Boars		0.91	0.81

In practical pig ration it is routine practice to add 0.5-1% limestone and 0.5% di-calcium phosphate/bone meal.

Calcium and Phosphorus Deficiency

Sows suffer from paralysis of the hind legs known as posterior paralysis. It occurs most commonly in high milk producing sows towards the end or just after the end of lactation. Ca and P content of pig milk are 2.5g and 1.7g/kg of milk, respectively. The gross efficiency of utilization of Ca and P is 47% and 50%, respectively.

Zinc Deficiency

There is 'parakeratosis' (Hyperkeratinization, thickening and reddening of skin). There is depression of appetite, reduced growth rate (subnormal growth), poor feed efficiency, characteristic skin lesion on legs and belly. In India, parakeratosis has been reported in some pig farms where high level of Ca i.e Ca level is twice the normal requirement (2% and above). A high level of Cu also causes the same problem to pigs. In this condition skin gives appearance of severe mange. High phytic phosphorus also aggravates the condition. Requirement for zinc is 50mg/kg diet.

Deficiency of Iron

Pigs are the only farm animals in which 'piglet anemia' is very common when they do not have access to soil or pasture and are housed in concrete floor. Piglets are generally born with low reserves of iron (50mg/ piglet) and due to rapid growth fall an easy prey to anemia if adequate iron supplements/ injections are not given.

Piglet anemia is characterized by labored spasmodic breathing (popularly known as thumps); hemoglobin level falls to 3-4mg/100ml of blood, poor growth rate, wrinkled skin, rough coat, lack of healthy pink color of visible mucous membrane. On post mortem, a dilated heart and edematous lung, watery blood is noted. Piglet anemia occurs 2-4 weeks after birth. The placental transfer of iron is so poor that the piglet is born with unusually small store of iron. The young piglets must retain 7-16mg iron/day to grow at a normal rate without becoming anemic whereas milk provides only 1 mg/day.

Prevention

An iron injection of Iron Dextran 100mg IM at 4th day and another at 14th day IM i.e. total 200mg. Saturated solution of ferrus sulphate (1.8%) @ 4ml daily orally. Painting the udder of the sow with a paste of ferrous sulphate causes intake of iron when the piglet suckles its mother and helps prevent thumps. As milk is deficient in iron and copper, care is taken during early growth to supply these nutrients. Requirement of copper is 6mg/kg diet.

Deficiency of Iodine (I)

There is hairless, dead piglets are born in deficient areas with enlarged thyroid. Iodine need of pig is 0.2 mg/kg diet which should be supplied in the form of iodized salt.

Deficiency of Manganese (Mn)

Deficiency shows poor growth of bones, enlarged hocks, muscular weakness, lameness with crooked and shortened legs, increase in back fat and irregular estrus cycle, reduced vitamin K induced blood clotting, and impaired glucose utilization. Requirement for Manganese is 10mg/kg diet.

Deficiency of selenium (Se)

This has a sparing effect on vitamin E. A deficiency may cause prolonged farrowing due to uterine atony, slow growth and impaired reproduction is common. Mulberry Heart Disease (MHD) or dietetic microangiopathy is seen due to gross morphological changes in the myocardium. There is haemmorhagic and necrotic lesion in the myocardium resulting into red mottling. Hepatosis diaetetica (HD) is seen in pigs between 3-15 weeks. This condition occurs in pigs when kept on a cereal (very low in Se) based ration for long period without supplementing Se. It is characterized by severe necrosis of liver

The Importance of Chromium (Cr)

For production of lean meat with high muscling and low fat, Cr is one of the most widely recommended minerals for high pork production. Cr @ 200 micro grams causes gain in weight, gain in muscle mass, and improvement in reproductive performance. Pregnant sows fed Cr had more total piglets born more live pigs and more pigs at 21 days than did sows fed an un-supplemented diet.

Requirement of Salt (NaCl)

As per the NRC, 0.5% common salt is recommended in rations of all classes and ages of pigs.

Requirements of minerals in general feed in pigs

Minerals	Pig
Ca	0.9%
P	0.6%
Mg	0.04%

[Table Contd.

Contd. Table]

Minerals	Pig
NaCl	0.25%
Na	0.10%
Cl	0.08%
K	0.23%
Fe	30-50ppm
Cu	3-6ppm
I	0.14ppm
Se	0.1-0.3ppm
Mn	10-12 ppm
Zn	80ppm

4.10. VITAMIN REQUIREMENTS

Deficiency of Vitamin A

Eye disorders such as xeropthalmia and blindness are common. In less severe case appetite is impaired and growth retarded, decrease in resistance to infection and decreased production of immunoglobulin. Blind deformed litters in vitamin A deficient pregnant sows, nervous disorder in pig like in coordination of gait, trembling, spasm, paralysis.

Deficiency in Vitamin D

In growing pigs there are symptoms like decreased feed intake and retarded growth, rickets, enlarged joints, broken bones, stiffness of joints, lameness and occasionally paralysis.

Deficiency of Vitamin E

There is nutritional muscular dystrophy in pigs. Heart muscle degeneration is seen which is known as 'Mulberry heart disease'. Increased mortality of newborn pigs, liver damage, edema, Metritis-Mastitis-Agalactia (MMA) syndrome.

Importance of Vitamin K

Vitamin K is synthesized in the intestine of the pigs by micro-organisms in adequate amount and hence has no practical importance.

Requirement of Vitamin B-Complex

Deficiency of vitamin B-complex may arise under practical conditions in pigs raised on feeds like cereal grains without much inclusion of green forage. Liberal supply of good quality legume fodder hay, dairy products will take care of vitamins of B-complex.

Deficiency of Thiamin (Vitamin B_1)

There is loss of appetite, reduced growth rate, vomiting, diarrhea, respiratory troubles, blue color due to inadequate oxygen, reproductive problems, disorders of nervous system, muscle weakness, insufficient energy utilization.

Deficiency of Riboflavin (Vitamin B_2)

Poor appetite, slow growth rate, digestive disorder like vomiting, diarrhea. There may be dermatitis leading to skin eruptions and loss of hair. Eye abnormalities like corneal opacity and cataract (opacity of eye lens) formation are also noted leading to impaired vision. There is nerve degeneration leading to in coordination, crooked and stiff legs in piglets. In case of sows, there is irregular estrus cycle and premature abortion are noted.

Deficiency of Pyridoxine (Vitamin B_6)

Anorexia, reduced growth rate, roughness of hair coat, diarrhea, poor vision, hepatic and cardiac damage, anemia (microcytic hypochromic), and demyelination of peripheral nerves leading to convulsion or epileptic fits. Goose step type of gait, fatty liver, dermatitis around eyes and snout leading to brown exudates around the eyes and snout is noted.

Deficiency of Cyanocobalamine (Vitamin B_{12})

Anorexia, slow growth, rough hair coat, in coordination of hind legs and a wobbly gait (due to muscular weakness and de-myelinated nerves), mild microcytic anemia is seen in young pigs. In adult pigs, there is sub-optimal growth, dermatitis, rough hair coat, reproductive failure and abortion.

Deficiency of Niacin

In pig this condition is known as 'Pig pellagra'. There is anorexia, poor growth, stomatitis, occasional vomiting, achlorohydria, diarrhea due to enteritis, dermatitis, poor skin condition, poor hair, alopecia, anemia, posterior paralysis.

Deficiency of Biotin

There is progressive dermatitis and dry scaly skin accompanied by alopecia, cracks in the feet, inflammation of hoof, lameness due to claw defects, and spasticity of hind legs. In growing pigs there is reduced growth rate, anemia, reduced FCR. In breeding sows, reduced reproductive performances like reduced litter size.

Deficiency of Pantothenic Acid

The growth rate is retarded. Scurvy skin, thin hair and loss of hair, ulceration of GIT, diarrhea, lactation failure, reproductive failure and a characteristic gait i.e. little bending of leg joints termed as 'Goose step' due to degeneration of myelin sheath. In this condition hind leg becomes stiff and staggers on walking. It may lead to paralysis. Scaliness of skin (dermatitis) is seen on shoulder and brown exudates around eyes.

Deficiency of Choline

Slow growth, fatty liver, abnormal gait, anemia, reproductive failure, reduced litter size, spraddle legs in newborn pigs. In adults, there is ataxia, fatty degeneration of liver and mortality.

Deficiency of Folic Acid

Growth and retardation, anemia, alopecia, poor reproduction and lactation.

Requirement of vitamin C: Vitamin C is synthesized in the body and hence not of practical importance.

Requirements of Vitamins in Feed in Pigs

Name of vitamin	Requirement/kg
A	2200IU/kg
D	220ICU/kg

[Table Contd.

Contd. Table]

Name of vitamin	Requirement/kg
E	11-22IU/kg
K	0.50mg/kg
B_1	1.5mg/kg
B_2	3-4mg/kg
B_6	1-2mg/kg
B_{12}	5-20mg/kg
Niacin	7-20mg/kg
Folacin	0.30mg/kg
Biotin	0.05mg/kg
Pantothenic acid	12mg/kg
Choline	0.5mg/kg

4.11. PRACTICAL FEEDING OF PIGS

Pigs grow faster than any other farm animals. In India, 70 kg body weight has been fixed for slaughtering pigs which is achieved in 6 months of age if fed scientifically but take 2 years in scavenging system. The average birth weight of a LWY piglet is 1.2 kg while an indigenous piglet weighs only 700 g. A LWY pig weighs 12 kg in a 8 weeks period and attain 65-70 kg in 6 months period which is the marketing age but in case European countries it is 90-110 kg.

Fermentation products i.e. volatile fatty acids may provide as much as 25-30% of maintenance requirement of pigs. The full-grown pig can handle 3-5 kg of leafy succulent green feeds (having 10-15% DM). In growing-finishing pigs 6-8% fibre may be used while the adults ration may contain 10-12%. Increase of crude fibre % beyond this maximum level will cause a depression of digestibility of GE by 3.5% for every 1% increase in dietary crude fibre. Pig prefers coarsely ground material instead of fine mixture. It is not necessary to soak or boil the grain feeds. Pigs can handle up to 40% of wheat or rice bran in growing-finishing rations without any depressing effect on growth rate and feed efficiency. Pigs are more susceptible to gossypol poisoning. Unsaturated fats produce oily carcass. Thus, only fats which will not produce oily carcass should be added to pig rations.

Feeding of Piglets

Milk is the main food of a piglet and source of energy is lactose sugar. Only glucose and lactose are effectively utilized by the piglets that are less than 7 days old. They can utilize fructose or sucrose only after 7 days of age. If a sow does not allow her piglets to suck her then milk replacers can be introduced. Use of milk replacer in piglets: Egg yolk-1 and cow milk- 1kg. Pigs less than 2-3 weeks of old have insufficient carbohydrate splitting enzymes in pancreas and intestine. The level of these enzymes rapidly increases after one or two weeks of birth, except lactase which goes on decreasing as age advances. So, it is advisable to offer starch or cereal based rations to the piglets only after 2 weeks. Pigs should be weaned by weight rather than by age. Creep feed is the feed given to the suckling piglets behind a barrier which allows only piglets to access to the feed. A good creep ration should contain about 24% CP and 3220 Kcal ME (NRC, 1988). Besides this the requirement for essential amino acids, minerals and vitamins should also be met. The crude fibre, total ash and acid insoluble ash in the creep ration should not be more than 5, 8, and 4%, respectively (BIS, 1986). The sow's milk is unable to support the growth of piglets which are exclusively kept on milk. Because, the sows milk is at peak at 3 weeks of age and after that it declines. So, Creep ration should be fed from 7-10 days onwards. Piglets achieved a growth rate of 300 gm per day with a feed efficiency of 2.9 kg per kg live weight. The creep mixtures are introduced at 7th day of age and are fed up to 56 days of age, by the time the piglets attain a body weight of 12 and 15 kg in Landrace and Yorkshire piglets, respectively. However, crossbred piglets attain a body weight of 8-12kg while the Indian native piglets attain a body weight of 5-8kg in 56 days of age.

Composition of Creep Ration (%)

Ingredients	I	II	III	IV	V	VI	VII
Ground Yellow Maize	46	40	40	60	40	30	20
Barley flour	-	-	-	-	-	10	10

[Table Contd.

Contd. Table]

Skimmed milk	-	10	-	-	10	10	10
GNC	20	10	20	20	30	30	30
Til cake	-	10	10	-	-	-	-
Fish meal	10	10	10	8	6	6	6
Wheat bran	10	10	10	10	-	-	-
Molasses	10	6	6	-	10	10	10
Brewer's yeast	2	2	2	-	2	2	2
Mineral mix.	1.5	1.5	1.5	1.5	1.5	1.5	1.5
Salt	0.5	0.5	0.5	0.5	0.5	0.5	0.5

Piglets are more prone to the piglet anemia. Generally, a dose of Cu and Fe may be given on 4th and 14th day. Injection of iron Dextran may be used intra-muscularly in such cases as discussed earlier in this chapter.

For feeding of piglets NRC (1988) suggested 3 (three) levels of protein i.e. for 1-5 kg 24% CP, for 5-10 kg 20% and for 10-20 kg body weight 18% CP. However, BIS (1986) has suggested only one level for protein i.e 20% in the starting phase.

Feeding of Growing-Finishing Pigs

After 56 days the piglets are switched from starter diet to grower diet. The crude fibre, total ash and acid insoluble ash in the grower ration should not be more than 6, 8, and 4%, respectively (BIS' 1986) The growing piglets are fed up to 50 kg body weight. NRC (1988) stated that up to 20 kg body weight requirement of CP is 18%, ME is 3250 Kcal/ kg diet; for 20-50 kg requirement of CP is 15%, ME is 3260 Kcal/ kg diet and for 50-110 kg body weight CP will be 13% and ME will be 3275 Kcal/ kg diet. According to the BIS (1986) only three types of feed for pigs are used and these are creep/ starter diet, pig growth meal and pig finisher or breeding meal.

Body weight (kg)	Time required	Stage of growth	CP% in the ration (BIS)
2-12 kg	Up to 56 days.	Starting	20% (Starter)
12-50 kg	2 months to 4 months	Growing	18% (Grower)
50-90 kg	4 months to 6/7 months	Finishing	16% (Finisher)

Growth Rations from 12-50 Kg Body Weight

After weaning at eight weeks of age, the average growth performance (gain in weight/day) of desi (local), crossbred and exotic piglets is in the range of 200-300, 250-400 and 300-600g, respectively. There are two types of grower rations: Cereal based grower ration and Non-cereal based grower rations.

Cereal Based Grower Rations

Maize/ barley have been fed as major component of the energy source. Here, the inclusion level of cereals may be more than 20%. The cereal based ration gives a growth rate of 450-550 gm was achieved up to 50 kg body weight with a feed efficiency varying from 3.8 to 4.2 kg per kg live weight gain.

Composition of Cereal Based Grower Rations (in %): (The ration should have more than 20% cereals)

Ingredients	I	II	III	IV	V	VI	VII
Yellow Maize	60	30	10	30	30	50	35
GNC	20	20	20	20	20	10	17
Wheat bran	10	40	60	40	-	32.5	40
Rice bran	-	-	-	-	40	-	-
Fish meal	6.5	6.5	6.5	6.5	6.5	5	6
Bone meal	1	1	1	1	1	-	-
chalk	1	1	1	1	1	-	-
Mineral mix.	1	1	1	1	1	2	1.5
salt	0.5	0.5	0.5	0.5	0.5	0.5	0.5

(**Source:** S.K. Ranjhan, 2001)

Low Cereal or Non-Cereal Based Growth Rations

Here, cereal grains have been replaced by the byproducts of concentrates (grains) either partially or completely in the growing-finishing pig. Here, the inclusion level of cereals may be less than 20%. On such rations the average daily gains were quite encouraging namely, between 350-450g/ day but efficiency of feed utilization was very poor, between 5-7 kg per kg live weight gain.

Composition of Low Cereal or Non-Cereal Based Grower Rations (in %): (The ration should have less than 20% cereals)

Ingredients	I	II	III	IV	V	VI	VII
Yellow Maize	10	-	-	10	-	10	-
Rice polish	-	-	70	60	35	30	-
GNC	20	20	20	20	10	10	7
Wheat bran	60	70	-	-	35	30	85
Til cake	-	-	-	-	10	10	-
Fish meal	7.5	7.5	7.5	7.5	7.5	7.5	6
Mineral mix.	2	2	2	2	2	2	1.5
salt	0.5	0.5	0.5	0.5	0.5	0.5	0.5

(**Source:** S.K. Ranjhan, 2001)

Finisher Ration

When the piglets have attained a body weight of about 50 kg, then they are switched on to the finisher ration till they attain a market weight of 70 kg. The limited work in India has shown that slaughter weight of 70 kg may be more economical than 90 kg body weight taken in some of the Asian countries. The crude fibre, total ash and acid insoluble ash in the finisher ration should not be more than 8, 8, and 4%, respectively (BIS'1986).

Lean Meat Production

For lean meat production, the restriction of feed intake is done when the pigs are about 4 months old or when they are at 50 kg body weight. Restriction of feed intake causes decrease in weight gain. Every 1% restriction in feed causes 0.53% decrease in back fat thickness.

Hardening Off

Adipose tissue is static and there is constant exchange of fatty acids between body fat and blood. Soft body fat of a pig can be converted to a hard fat if we change the dietary fat, in fact fatty acid. Feed containing high amount of saturated fatty acids can be fed to the pig when the pig is on a feed that contains a high amount of unsaturated fatty acid that produces a soft body fat. Now, the soft body fat will be converted to hard body fat and the process is called 'Hardening off'.

Composition of Finisher Rations for Pigs (in %):

Ingredients	I	II	III	IV	V
Yellow Maize	40	40	40	-	-
Rice polish	-	30	10	-	25
GNC	12	12	12	17	17
Wheat bran	30	-	20	75	50
Til cake	10	10	10	-	-
Fish meal	5.5	5.5	5.5	5.5	5.5
Mineral mix.	2	2	2	2	2
salt	0.5	0.5	0.5	0.5	0.5

(**Source:** S.K. Ranjhan, 2001)

Feeding Schedule of Growing-Finishing Piglets in general

Age	Daily feed intake (kg)	Feed consumption per pig in a month (in kg)	Cumulative feed consumption per pig in a month (in kg)	Probable live Weight (kg)	Gain in live weight (kg)	Daily gain in weight (g)
Birth-1 month	Sow's milk	Small amount creep feed	-	6 kg	5 kg (say live birth wt. is 1 kg)	166g
1-2 months	0.5	15	15	12	6	200g
2-3 months	1	30	45	25	13	433
3-4 months	1.25	37.5	82.5	40	15	500
4-5 months	2	60	142.5	60	20	666
5-6 months	2.5	75	217.5	70-75	10-15	350-500
Total		**217.5 kg**				

Here F.C.R. = **2.9-3.1:1**

BIS Specification for Pig Feeding (1986) Reaffirmed in 2001

Characteristic	Creep/Starter feed	Pig Grower meal	Finishing/ Breeding meal
Moisture % max.	11	11	11
Crude protein % min.	20	18	16
Metabolizable Energy (Kcal/kg),min	3360	3170	3170
DE(Kcal/kg), min	3500	3300	3300
Crude fibre % max.	5	6	8
Total Ash%, max.	8	8	8
AIA % max.	4	4	4
Ether extract %	2	2	2
Vitamin A(IU/kg)	1700	1300	1300
Riboflavin, mg/ kg	3	2.4	2.2
Niacin, mg/kg	17	14	10
Pantothenic acid, mg/kg	11	10	10
Vitamin B_{12} microgram/kg	15	11	11
Vitamin D (IU/kg)	190	180	130
Calcium %	0.6	0.6	0.6
Avl. Phosphorus %	0.6	0.4	0.5
Iron, mg/kg,min	100	90	80
Copper, mg/kg, min	8	6	6
Manganese, mg/kg,min	30	30	20
Zinc,mg/kg,min	50	50	50
Common Salt, max	0.5	0.5	0.5

NRC Specification for Pig Feeding (1988)

Characteristics	Body weight				
	1-5 kg	5-10 kg	10-20 kg	20-50 kg	50-110 kg
Moisture % max.	11	11	11		11
Crude protein %min.(1988)	24	20	18	15	13
CP% (as per NRC'98) 80kg) & 13.2 (80-120kg)	26	23.7	20.9	18	15.5(50-
Metabolizable energy (Kcal/kg),min	3220	3240	3250	3260	3275
DE (Kcal/kg), min	3400	3400	3400	3400	3400
Expected wt. gain g/day	200	250	450	700	820
Expected feed intake g/day	250	460	950	1900	3100
Expected efficiency feed/gain	1.25	1.84	2.11	2.71	3.71
Lysine %	1.4	1.15	0.95	0.75	0.60
Vitamin A(IU/kg)	2200	2200	1750	1300	1300
Riboflavin, mg/ kg	4	3.5	3	2.5	2
Niacin, mg/kg	20	15	12.5	10	7
Pantothenic acid, mg/kg	12	10	9	8	7
Vitamin B_{12} microgram/kg	20	17.5	15	10	5
Vitamin D (IU/kg)	220	220	220	150	150
Calcium %	0.9	0.8	0.7	0.6	0.5
Avl. Phosphorus %	0.55	0.40	0.32	0.23	0.15
Iron, mg/kg,min	100	100	80	60	40
Copper, mg/kg,min	6	6	5	4	2
Manganese, mg/kg,min.	4	4	3	2	2
Zinc, mg/kg,min	100	100	80	60	50
Selenium, mg/kg	0.3	0.3	0.25	0.15	0.10

Ideal Amina Acids Profile for Pig Feeds

Lysine	100
Methonine + Cystine	75
Threonine**	65
Valine	80
Isoleucine	67
Arginine	105
Tryptophan	17
Histidine	40
Leucine	105
Phenylalanine + Tyrosine	105

Nutrient Requirements of Breeding and Lactating Pig (NRC, 1988)

Intake Level	Breeding Gilt and Sows (110-160kg)	Lactating Gilt and Sows (140-200kg)	Boars (110-180kg)
ME, Kcal/kg	3210	3210	3210
DE, Kcal/kg	3400	3400	3400
CP%	12%	13%	12%
Calcium%	0.75	0.6	0.75
Phosphorus%	0.50	0.40	0.50

Feeding to Breeding Sows and Boar

Sows and boars that are underfed may become too thin to reproduce normally. Similarly, those that are overfed may lack libido or physical stamina due to obesity. To prevent overfeeding, boars and gilts that are selected for breeding should be fed at restricted level from 6 months to breeding age i.e up to 8-9 months. Males and females used for breeding must be fed nutritionally adequate diets in order to maintain body weight and reproductive performance. Careful attention is required in ensuring adequate intake of Ca, P and other inorganic elements for better sound health.

Flushing of Sows

After weaning the piglets at 8 weeks of age, the sows should be flushed. It means putting the sow on a more nutritious ration (feeding liberally) and then mated during the first heat period. Due to flushing a live weight gain of 500g per sow per day is desired during 1-2 weeks prior to breeding. 0.5-0.7 kg ration per head per day for 1-2 weeks prior to breeding season is required. A well- balanced high protein diet 1-2 weeks prior to breeding is required for flushing to obtain greater litter size and litter weight. Other advantages are: It improves physical condition of the female, shows prominent heat symptoms, increases ovulation rate, shortens period between weaning to successful conception, more uniform litter size and minimizes embryonic death, prompt post weaning estrus, etc.

Feeding of Sows During Gestation

Pregnant sow will consume far more feed voluntarily than is required and will become obese if intake is not restricted to a daily intake of 1.8-2.3 kg. Hence, during gestation period, gilts and sows are fed on restricted feeding (2-2.5 kg) without getting them over-fat. 2-2.5 kg feed may be sufficient for a pregnant sow but for gilt it may be 3 kg. Rest of the feeding may be done with good quality succulent feed. Berseem, Lucerne, cowpea etc. may be fed (*25-30% of ration*) between 5-6 kg/head/day. However, during the last two weeks before the furrowing daily feed intake should be increased to 2.5-3.25 kg in order to accommodate the surge in weight of the foetus in the late pregnancy. An average daily weight gain of 250-300 g ($0.25\text{-}0.30 \times 114 = 29\text{-}34$kg) throughout the gestation period is satisfactory. The sow in gestation should gain 25-35 kg compared to gilt 30-40 kg. A proper feeding of a pregnant sow will result in:

- A large litter in both number and weight.
- Healthier piglets at birth.
- Enough milk yield by the sow.
- More and heavier pigs weaned.
- Reduction in stillbirth, abnormalities and stunted growth.

Requirements for vitamins are higher for the breeding animals. The vitamin supplements may be doubled while feeding the pregnant sows/gilt.

A pregnant sow should not be overfed as it may cause various problems. A higher amount of feed will cause embryonic mortality. Milk of over fat sows contains more fat and less protein. Over fat sows do not like to eat after farrowing which reduces milk yield and ultimately affects the piglets. Fat causes problem like dystokia during farrowing time.

Feeding of Sows During and After Farrowing

Three or four days before farrowing, the ration should be reduced to half and made more laxative through the introduction of more wheat bran/rice bran. Half of ration may be substituted by wheat/rice bran to prevent constipation. No feed should be offered on the day of farrowing but plenty of lukewarm water for drinking purpose must be given. On the next day 1-1.5 kg of laxative ration should be given. On successive days, the amount fed can be increased by 200-300g daily through gradual introduction of the regular lactation ration so that a sow is on full feed in about 7 days after farrowing.

Feeding of Sows in Lactation

The requirements of the lactating sows are greater than those during the gestation period. A sow in full milk production requires 4 to 6 kg of feed per day depending upon body weight, its condition, milking capacity and size of litter. Thumb rule for feeding for exotic/crossbred sow: Provide 3-3.5 kg as maintenance ration plus 0.3-0.5 kg for each piglet, the sow is nursing. In case indigenous sows it is 2.5 kg for maintenance and 0.25 kg feed per piglet till the piglets are weaned. A lactating sow having 8-10 litter size requires 15-25 liters of water per day.

Feeding of Boar

In case of breeding boar, meals having composition of finisher ration (16% crude protein as per BIS or 13% CP as per NRC) may be fed at the rate of 2.5 kg/ head/ day. The rest of the ration may be provided as green roughage. Boar should neither be underfed nor fattened but kept in good thrifty condition. Boar should not be too fat as it weakens leg and breeding ability. They should be given regular exercise to keep in good health. Boar should be fed 2-3 hours before they are expected to breed. Always provide fresh, clean and adequate quantity of water.

Organic selenium may be supplied for improved sperm concentration and quantity. It is recommended to supply 50 percent of total needs (up to 0.5 ppm). Vitamin E is for enhancing sperm quality and the recommended dietary specification up to 250 IU/kg. Zinc is required for enhanced sperm quality. Its recommended dietary specification is up to 200 mg/kg. This also helps in maintaining hoof health, which is very important for large-frame and heavy boars, especially those used for natural service. Biotin is necessary for better feet health. Recommended dietary specification is up to 0.5 mg/kg. Vitamin C is supplemented to ameliorate heat stress. Recommended dietary specification is up to 500 mg/kg. It should be used along with other measures to reduce heat stress. Betaine is also supplemented to ameliorate heat stress. Recommended dosage varies from 1 to 2 or more kg per metric ton of feed. It is a powerful cellular hydrant, and access to abundant water is thus important. Chromium tripicolinate is for enhancing performance. Recommended dietary specification is up to 200 ppb.

Nutrient Requirements for Breeding Pigs (NRC, 1988)

Characteristics	Pregnancy	Lactation	Boar
Feed Consumed (kg)	1.9-2.4	**5.3-5.9**	2.2-2.6
Crude protein % min.(NRC' 1988)	13	13.5	13
DE (Kcal/kg), min	3000	3240	3000
Lysine %	0.48	0.8	0.65
Meth+cys %	0.25	0.40	0.33
Threonine%	0.29	0.48	0.39
Isoleucine%	0.26	0.44	0.36
Leucine%	0.48	0.90	0.65
Tryptophan%	0.07	0.14	0.10
Vitamin A (IU/kg)	6000	6000	6000
Riboflavin, mg/ kg	3	3	3
Niacin, mg/kg	15	15	15
Pantothenic acid, mg/kg	10	10	10
Vitamin B_{12} microgram/kg	15	15	15

[Table Contd.

Contd. Table]

Characteristics	Pregnancy	Lactation	Boar
Vitamin D3 (IU/kg)	750	750	750
Calcium %	0.85	0.88	0.85
Avl. Phosphorus %	0.65	0.68	0.65
Iron, mg/kg,min	60	60	60
Copper, mg/kg,min	5	6	5
Manganese, mg/kg,min.	15	16	15
Zinc, mg/kg,min	50	50	100
Selenium, mg/kg	0.15	0.15	0.15
Iodine mg/kg	0.5	0.5	0.5
Salt (g/kg)	0.35	0.35	0.35
Essential fatty acid (g/kg)	7	7	7

4.12. FEEDING OF ANTIBIOTIC IN PIGS

- Antibiotic reduces incidence of scouring unthriftiness in pigs.
- The good effect of feeding antibiotic feed supplement is seen in animals given all vegetable protein diets than those getting animal protein diets.
- The growth rate is better when antibiotic is used in an unhygienic condition than when used in hygienic condition.
- Higher the nutritive value of ration, the less would be the improvement in growth rate on antibiotic feeding.
- A mixture of two or more antibiotic is no more effective than the single effective antibiotic.
- The greatest benefit of feeding antibiotic is seen during the early growth period (between weaning to 50 kg body wt.). The effect diminishes with ages.
- The runty pigs show better performance with antibiotic feeding. Number of runts pigs also decreases thereby make more uniform crop of pig.
- Due to antibiotic feeding the increase in growth rate may vary between 10-20% and reducing the feed intake by about 2.5%.

- Antibiotic improves the efficiency of feed utilization by 5-8%.
- The dose rate is 15-50 mg/ kg feed and no advantage is seen in exceeding these levels. e.g. Virginamycin, Bacitracin, Flavomycin, Tylosin, etc.
- In case of sows: There is shortened weaning to mating interval, there is increased litter size, improved quality of milk of sows (milk fat and solid not fat content increased), increased weaning weight of 1kg per piglet, less body weight loss during lactation period.

4.13.USE OF PROBIOTIC IN PIGS

It is very useful in very young and growing pigs. It can control incidence of pig diarrhea. It improves feed efficiency and weight gain. Increased feed intake, daily increase in weight gain is noted in probiotic fed growing pigs. Sows receiving probiotics weaned larger, heavier litters. Incidence of *E. coli* scours in piglets is also reduced.

Live Yeast Culture for Pig and Its Effect

- Better feed efficiency.
- Increased body weight gain.
- Better intestinal microbial balance.

Copper Supplements

It is used in grower pig ration as a growth promoter. Copper sulphate has bactericidal property. So, the copper effect is similar to antibiotics. Dose rate is 0.01% of the diet in fattening pigs to improve rate of gain and feed efficiency.

Enzymes as Feed Additives

Maize and soyabean meal are the most preferred feeds for poultry and pig. Due to high cost of the above ingredients like wheat, oats, barley, etc. and sunflower seed meal, rapeseed meal, etc are being used nowadays. But unfortunately, the digestibility of these ingredients is less. The presence of non-starch polysaccharides (NSP) in the cereal grains are cellulose, araboxylans and beta-glucans. Some of the important commercial enzymes are listed below.

Enzymes	Substrates
Xylanase	Xylans
Arabinase	Arabinans
Mannase	Mannans
Glucanase	Beta-Glucans
Pectinase	Pectins
Cellulase	Cellulose
Hemicellulase	Hemicellulose
Phytase	Phytate phosphorus
Protease	Proteins
Lipase	Lipids
Galactanase	Galactants

Use of Buffer in Pigs

Sodium bicarbonate ($NaHCO_3$) is used in the pig ration @ 10kg/ ton of food (1%) to reduce the chances of gastric ulcers.

Chemical Composition and Nutritive Value of Common Pig Feed Ingredients

Ingredients	ME Kcal/kg	DE Kcal/kg	CP	CF	EE	Ca	P	Lysine	Methio-nine
Yellow maize	3670	3900	9	2.2	3.8	0.02	0.28	0.22	0.18
Sorghum white	3170	3170	10	2.3	2.8	0.03	0.28	0.21	0.16
Bajra	2350		11.5	3.5	4.3	0.06	0.33	0.43	0.23
Broken rice	4800		8.5	10.6	1.9	0.08	0.39	0.24	0.16
Tapioca	3150	3540	3.5	4.3					
Wheat	3550	3112	10	2.4	1.8	0.05	0.31	0.30	0.16
Oil	8000	8750	-	-	-	-	-	-	-
DORB	2550	2400	13.5	14	0.6	0.07	1.5	0.6	0.25
Rice polish	3050	3500	12	8	15.1	0.08	1.3	0.5	0.22
Wheat bran	2760	2760	15.7	11	3	0.14	1.15	0.59	0.23
Sunflower cake	2550	3350	27	28	1.1	0.37	1	1.13	0.58
SBM	2680	3490	45	6.6	0.8	0.29	0.65	2.7	0.65

[Table Contd.

Contd. Table]

Ingredients	ME Kcal/kg	DE Kcal/kg	CP	CF	EE	Ca	P	Lysine	Methio-nine
GNC-SE	3000	3415	42	13	1	0.2	0.63	1.6	0.45
GNC-EXP	3580		40	13	7.3	0.16	0.56	1.5	0.42
Rapeseed cake	3710		35	11	1.4	0.72	1.12	1.7	0.65
Fish meal	2600	2700	42	1	5	3.73	2.43	3.2	1.1
Meat meal	-		45	8.7	7.1	8.27	4.1	2.5	0.65
Blood meal	-		88	-	1.8	-	-	7	1.14

4.14. FEED EFFICIENCY IN PIG NUTRITION

The amount of feed required for an animal to make a unit gain in weight is called feed efficiency. Pig is the best converter of animal feed to human food among the farm animals.

Factors Affecting Feed Efficiency

1. **Quality of Diet:** An optimum energy: protein ratio and balanced vitamins and minerals lead to increased feed efficiency.
2. **Environmental Temperature**: At low temperature maintenance energy requirement is more. So, there is reduced energy supply for growth and ultimately there is decreased feed efficiency.
3. **Age of Animal**: In young animals tissue growth rate is very high (cell multiplication is very fast). They need less feed per unit of weight gain than older animals. FCR in growing LWY pig is 3.13-3.35, whereas finishing LWY pigs have 3.82-4.13.
4. **Rate of Growth:** Faster gaining animals have an increased feed efficiency than the slow growing animals.
5. **Amount of Fat in the Body**: In older animals, % of fat increases. If excess fat is decreased then feed efficiency increases because fat occupies 90% of DM where as lean tissue occupies only 30% of DM.
6. **Heritability:** Feed efficiency in pig is 30-40% heritable.

Requirement of Drinking Water for Pigs in general

Classification	Age (in months)	Drinking water(L)/pig/day
Growing pigs	2-3	3-3.5
	3-5	6
	5-6	7-8
Pregnant gilt/sow	First 2 months of pregnancy	10-15
	Last 2 months of pregnancy	15-20
Lactating sow	Litter size: 8-10	20-25
	Litter size: more than 10	25-30
Mature boar		20-25

Example of Different Types of Pig Ration (As Per BIS, 1986)

Feed ingredients	Creep/Starter	Grower	Finisher
Maize	60	55	57
DORB	8	23	28
GNC (exp.)	20	13	8
Fish meal	10	7	5
Mineral mixture with salt	2	2	2
Vitamin AB_2D_3	20g/100kg	20g/100 kg	20g/100kg
Antibiotic	20g/100kg	20g/100 kg	20g/100kg

Calorific Value of Indian Feeds for Pig (Rostock Equation)

Feed stuff	DE kcal/kg feed	ME kcal/kg feed
Maize	4.11	3.67
Oat	3.44	3.07
Soybean	4.28	3.83
Wheat	3.99	3.55
GNC	4.04	3.58
Sarson cake	3.95	3.53
Til cake	4.33	3.84

[Table Contd.

Contd. Table]

Feed stuff	DE kcal/kg feed	ME kcal/kg feed
Rice bran (DO)	2.82	2.55
Rice bran	3.34	3.05
Wheat bran	3.35	2.97
Tapioca	3.54	3.15

Inclusion Level of Some Unconventional Feed Resources for Pigs

Feed stuff	Level of inclusion
Tapioca	25-30%
De-oiled rice bran	25%
Sweet potatoes (raw)	25-40%
Sweet potatoes (dehydrated)	25-30%
Potatoes (raw)	20-40%
Potatoes (dehydrated)	25%
Jackfruit	50%
Cottonseed meal	8-9%
Coconut meal	10-20%
Linseed meal	5%
Sunflower meal	8%
Safflower meal	5-10%
Til cake	10%

CHAPTER 5

PIG REPRODUCTION

Dr. Dipanjan Chakraborty

5.1. INTRODUCTION

Domestic pigs are generally polyestrous, while European wild sows usually exhibit estrus twice a year depending on the season. The estrous cycle in domestic pigs is for 21 days and may vary depending on the breed and seasonal differences. The estrous cycle comprises four stages:

1. Proestrus which lasts for 1-3 days and is more prominent in young females with vulvar hyperemia. The female does not allow the male to mount her.
2. Estrus lasts for approximately 2-3 days, and ovulation occurs in the last third of estrus, 36-44 hours after its onset. The vulva of pigs during estrus is hyperemic and edematous. There is also vaginal discharge. Behavioural changes in female pigs during estrus include an increase in activity, tracking of moving objects, shouting, and irritability. Female pigs also exhibit mounting and frequent urination in the presence of males. The mating position which is also known as lordosis (remaining still and standing with four feet in extension) and upright ears are the other signs of estrus. The mating posture, with a straight back or waist with both front feet touching, and the smell of male pheromones can also be visually observed.
3. Metestrus lasts for about 2 days and is very similar to diestrus.
4. Diestrus lasts for approximately 14 days, and during this time, females strongly reject the mating by the males.

DOI: 10.1201/9781003749844-5

During the late estrus or early metestrus stages, a whitish mucous discharge containing mucus, cellular debris and leucocytes may be observed in the vulva of sows. The vaginal epithelium in sows thickens cyclically when in estrus, forming 8 to 15 cell layers. This thickening decreases to 4-5 layers in diestrus and 2- 3 layers during pregnancy. Therefore, a vaginal biopsy can be a useful tool for pregnancy diagnosis in pigs, particularly in field conditions.

The ovulation rate is low during the first cycle but increases with subsequent cycles. Crossbreeding, a high-energy diet, and mating after the third heat can increase the ovulation rate and result in larger litters. Domestic sows breed all year round while wild pigs breed seasonally. Reducing daylight between May and August can shorten the time between weaning and ovulation in domestic sows. Ovulation after weaning generally occurs within 4-6 days when piglets are weaned at 5-6 weeks.

5.2. PUBERTY

Female pigs, or gilts, reach puberty at around 7 months and their body weight is around 100-110 kg. However, the age of puberty can vary depending on several factors, including genetics, environment, social surroundings, and nutrition. Several factors affect the age of puberty in pigs, which are as follows:

1. **Age and weight:** Extreme weakness can delay puberty and also reproductive performance. Induced early puberty can reduce birth rate, live birth rate, and weaning mortality in piglets. However, these effects are only seen in pigs mated after 270 days.
2. **Stock:** Mixed breeds reach puberty earlier than pure breeds.
3. **Genotype and body condition score:** it is observed that to attain puberty a minimum back fat thickness should be 6 mm.
4. **Feeding:** Short-term changes in feed intake do not affect live weight or body condition, but they can adversely affect reproductive function.
5. **Housing:** Overcrowding can delay puberty as well as estrus.
6. **Season:** Heat stress can also delay puberty. therefore, sows show more estrus signs in the winter season compared to summer.

7. **Environmental atmosphere:** Poor ambient air quality can delay puberty in gilts.

5.3. ENDOCRINOLOGY OF ESTROUS CYCLES

The duration of the estrous cycle in pigs is approximately 21 days and can be divided into three phases: the follicular phase (proestrus), the ovulatory phase (estrus), and the luteal phase (metestrus and diestrus).

The follicular phase begins with luteal regression and ends at ovulation. During this phase, the follicles which ovulate grow rapidly due to increased LH release. When the estradiol levels peak, it triggers the LH surge as a result the pig showing estrous behavior. Sexual receptivity generally lasts 1-4 days. Sows usually ovulate 36-44 hours after the onset of estrus generally toward the end of the estrus. Natural ovulation in sows lasts about 2 hours, but some studies suggest it may last up to 9 hours.

During the ovulatory phase, the LH surge reduces the production of estradiol and shifts towards progesterone production. The oocyte will resume meiosis in response to the gonadotropin surge, and meiosis will be completed after sperm penetration. The role of follicle-stimulating hormone (FSH) during the ovulatory phase has not been as well-characterized as that of LH. A surge in FSH has been reported to coincide with the LH surge, but the post-ovulatory surge in FSH is more distinct and appears to be responsible for activating small follicles for potential ovulatory growth during later cycles.

The luteal phase is the period of formation of corpora lutea and the production of progesterone hormone. Serum progesterone increases between 2 and 4 days after estrus. Corpora lutea are formed from the cellular remnants of follicles after ovulation, which is called corpora hemorrhagica. The corpora lutea of pigs contain cells of follicular granulosa (large luteal cells) and thecal (small luteal cells) cells.

5.4. ESTRUS DETECTION

In the pig farms, estrus is generally monitored twice a day, at 7:00 am and 5:00 pm. However, sometimes it is only monitored once a day, either 30 minutes before or after feed intake, due to the farm's work routine. The first method increases the chances of determining the onset of estrus and achieving fertilization after artificial insemination.

5.5. ESTRUS SYNCHRONIZATION AND RELATED EVENTS

In cyclic pigs, prostaglandin F2α (PGF2α) or its analogs are not effective in regressing the corpus luteum up to day 12 of the sexual cycle. Therefore, PGF2α cannot be used for synchronization until the 12th day of the estrus cycle. Gonadotropin treatment is a synchronization method used in adult pigs and sows at weaning to prevent a delayed return to estrus. An alternative dose and length of protocol of PGF2α can be used. To shorten the weaning-estrus and anestrus interval, 2 mL of Lutalyse (dinoprost tromethamine) can be injected within the first 24 hours after birth, and 5 mL of PG600 can be applied during the subsequent weaning treatment. Estrus is usually observed within 3-7 days after this practice. Sows can also be synchronized by the oral administration of progestin for 14-18 days. After the cessation of treatment, 95% of treated sows display estrus within 4-9 days. A methylcellulose gel vehicle used intravaginally after birth to stimulate estrus has also been developed. When 100 g of triptorelin is combined with this gel 96 hours after weaning, the length of estrus, the estrus-to-ovulation interval, and the weaning-to-ovulation interval are all reduced.

In pigs, mating usually lasts between 5 to 20 minutes. After ejaculation, the sow is filled with about 200 to 250 mL of sperm in the cervix and uterus. The sow's behavior changes after mating and it doesn't allow the boar to mate with her again for the next 1 to 2 hours. Boars can mate with a female up to 8 times, while sows need to be mated at least 3 times to increase the possibility of fertilization.

In pigs, the **maternal recognition** of pregnancy is achieved by the presence of at least 2 embryos in each horn. If there are less than 2 embryos, PGF2α is synthesized and the pregnancy is terminated with luteolysis.

The pig placenta is an epitheliochorial/ diffuse type. After ovulation placenta 1st attached to the uterus around the 12th day, and sufficiently developed around 18-20 day.

5.6. PREGNANCY

The gestation period of domesticated pigs usually lasts for an average of 114 days. However, in wild pigs, it can last up to 124 to 140 days. It's important to note that not all embryos are the same size and quality. Some of them may be defective or less developed. During the first ten days, many embryos can be absorbed, and up to the 20th day, the total loss of embryos can be about 40%. This is a crucial factor in determining litter size.

Pregnancy recognition -can be done through clinical or laboratory methods

Clinical methods include ultrasonography, which can be done either transcutaneously or transrectally. Transcutaneous ultrasound is applied to the sow's right lateral ventral abdomen, to keep the animal calm, and embryos can be detected as early as 19-20 days using a 5-MHz probe. On the other hand, transrectal ultrasound requires the removal of the animal's faeces by hand before inserting a probe into the rectum. Embryos can be detected as early as 18 days using a 7.5-MHz probe. However, diagnosis by ultrasound is not practical before the 20th day of pregnancy.

Laboratory methods include vaginal biopsy, which counts the number of epithelium layers in the vaginal mucosa, and plasma progesterone (P4) level measurement. A P4 level of $\geq$ 7.5 ng/mL is an indicator of pregnancy, with an accuracy rate of around 96%. Measurement of plasma estrone sulfate levels are another laboratory method, which can be done at 24-30 days of pregnancy when a high concentration of it is present in the blood plasma. Estrone sulfate decreases to a low concentration at 35-45 days of pregnancy, while a second increase occurs at 70-80 days and continues until farrowing. This method requires less blood because a small amount of blood taken from an ear vein is enough.

5.7. TERMINATION OF PREGNANCY

It is possible to use PGF2α or synthetic analogues to terminate the pregnancy at any stage. When 10 mg of PGF2α or 175 mg of cloprostenol is injected into the muscle on the 112th or 113th day of pregnancy, farrowing will take place within 20-36 hours. Stimulation of birth can be achieved with the use of a single daily dose of 75-100 mg corticosteroid on days 101-104, leading to birth on day 109. However, this method is not practical as it is expensive and reduces the number of live-birth piglets. A single dose of cloprostenol is usually not enough to stimulate the release of prostaglandin. however, it can, be used in combination with oxytocin. By using 0.18 mg of cloprostenol and 40 IU of oxytocin, farrowing can be induced within 20-35 hours.

5.8. PARTURITION

Parturition is controlled by a complex set of physical and hormonal events. Pregnant sows become more active and avoid other pigs before giving birth. Swelling of the vulva and mammary glands occurs a few days before birth. About 60-75% of pig farrowing occurs at night, with the second stage of labor lasting between 150-180 minutes. Sows lie down while strong contractions help to expel piglets, with an average of 16 minutes between successive piglet expulsions. Enhancing reproductive performance on swine farms is challenging and requires sound management practices and rigorous genetic selection programs.

5.9. DIFFERENT TYPES OF NONINFECTIOUS CAUSES OF INFERTILITY IN PIGS

Anoestrus is when a sow does not exhibit estrus behaviour. It is a natural occurrence before puberty, during pregnancy, and lactation. Delayed estrus is included in this category. Minimizing even "normal" periods of anoestrus is crucial to reducing non-productive days. The average weaning to estrus interval is 4 to 6 days, and any delay results in a loss of production. Anoestrus is a common reproductive disorder in sows and gilts, with a higher prevalence during the summer months. Factors such as parity, geographical location, and boar presence may also influence its incidence.

Suboestrus is a condition in cyclic animals where there are no external signs of estrus, despite the presence of corpora lutea. To ensure accurate estrus detection, boar presence and record-keeping are necessary. Sows must be exposed to a boar for 20 minutes daily from weaning and observed for signs of estrus. Early estrus may be missed if detection is delayed. Estrus detection is challenging in gilts, with detection rates as low as 50-60%.

Tests for ovarian function are necessary to differentiate between inactive, suboestrous and pregnant ovaries. Progesterone measurement can help identify changes in ovarian activity. The presence of the boar is key in stimulating estrus in sows and gilts. Exposure to mature boars is the most important feature of the post-weaning environment to ensure a rapid return to estrus and ovulation.

During a post-mortem examination, the reproductive tracts of sows culled from the herd under investigation can be analysed to assess ovarian function. This type of study can identify various abnormalities, such as:

1. Acyclic ovaries: These are inactive ovaries that have small follicles (<5 mm diameter) and no corpora lutea.
2. Cystic ovaries: These are ovaries containing multiple cysts. The cysts can be large (usually <14 mm in diameter) and generally contain luteal tissue that produces progesterone. While some cysts may regress, others can persist and inhibit estrus. Alternatively, the cysts may be small, resulting in sows having markedly irregular cycle lengths and exhibiting intense signs of estrus (nymphomania). Finally, single cysts are rare and tend to be incidental findings at post-mortem, and they typically do not affect sow fertility.

Real-time ultrasound is a non-invasive method to observe changes in the ovaries of live animals. It requires expensive equipment and expertise but it is useful for diagnosing infertility and researching ovarian activity.

Anoestrus is a condition where the ovaries of a female animal are inactive and it is of economic significance in animal farming. All contributing factors must be identified and remedied promptly. In severe cases, hormone therapy may be considered to restore cyclical activity. There are various hormone preparations available, but a combination of equine chorionic gonadotrophin (eCG) and human chorionic gonadotrophin (hCG) is still probably the best option. A injecting containing 200 IU hCG and 400 IU of eCG (PG600, Intervet) resulted in showing signs of estrus within 3-8 days in most of the sows.

5.10. CONCEPTION FAILURE

An incidence of returns higher than 10% is unacceptable as it indicates conception failure. The conception rate must be at least 90% and failure to achieve this is due to one or more factors, assuming ovulation has taken place. Conception failure leads to either total or partial conception failure, which is not acceptable.

Timing is crucial for successful breeding of sows. A single mating during the right stage of estrus can lead to a high rate of fertilized ova, embryonic survival and larger litter size at birth. However, identifying the right moment is difficult. To maximize the chances of conception, a service routine should ensure that the uterus contains viable spermatozoa before and during the time of ovulation. The ideal service management regimen would ensure that each sow is served on the day of onset of standing estrus and again 18-24 hours later.

Supervision is necessary to ensure proper service and intromission. Service pen design, especially the floor surface, affects quality. Young boars need training and close supervision when mounting sows or gilts.

Semen quality is an important aspect of fertility that should not be overlooked. It can be affected by various factors, and if affected, conception rates may also be affected. Regular returns to service may be due to a management problem rather than an infectious disease. Sufficient time should be spent on estrus detection.

CHAPTER 6

ARTIFICIAL INSEMINATION IN SWINE

Dr. Amit Kumar Nath

6.1. HISTORY OF SWINE AI

The AI of swine was initiated by Ivanov in Russia in the early 1900s. The technique was quickly spread to the United States (McKenzie, 1931), Western Europe (Polge, 1956) and Japan (Niwa, 1958). The first diluents as a method to store boar semen were developed in Russia (Ivanov, 1922, Milovanov 1938) and in the United States (Philips and Lardy, 1940) for storage of boar semen at 7-12^{0}C. However, Ito *et al.*, (1948) recommended storage at 15-20°C which is used at present. The mostly used semen extender is Beltsvile Thawing Solution (BTS) developed by laboratories of the United States, Department of Agriculture by Pursel and Johnsons in 1975. A non-surgical technique for semen deposition into the uterus was first described in 1959 by Hancock.

6.2. COLLECTION OF SEMEN, DILUTION AND PROCESSING

Male Reproductive Tract of Swine

The testis in the boar is oval in shape, 10 to 15 cm. long and 5 to 9 cm. in diameter. The testis of mature boars weighs between 500 to 800 gm. The tail of the epididymis is dorsal, the body cranial and the head of the epididymis ventral. The tail of the epididymis is quite large. The ductus deferens lies cranial and medial to the testis. Vesicular glands are 5 to 8 cm wide and 5 to 8 cm in thickness covering bladder caudally and extend into the abdominal cavity. The prostate gland is located dorsal to the bladder and is covered by the vesicular glands. Bulbourethral glands are large,

DOI: 10.1201/9781003749844-6

dense and cylindrical in shape about 12 to 18 cm in length, 2.5 to 5 cm in diameter and lie on either side of the pelvic urethra. The penis is characterized by the presence of prescrotal sigmoid flexure and small glans cranially which is spirally twisted counter clockwise. The penis is 45 to 55 cm in length and at the time of copulation about 20 to 35 cm protrudes beyond the preputial opening. The prepuce of the boar has a small orifice with long bristle like hairs. The prepuce is narrow caudally and wide cranially.

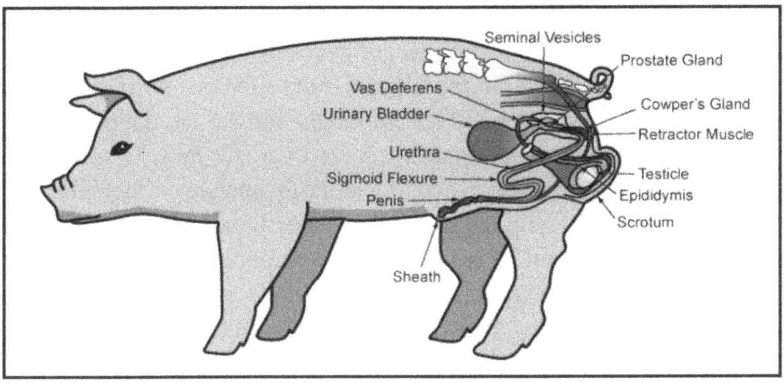

Fig. 1: Reproductive tract of boar (photo source: https:// www.missouricareereducation. org/doc/animalrepro/Student Resource. pdf)

Collection of Semen

Estrus female is not required for collecting semen. Semen is collected twice per week. The duration of ejaculation period is 5-10 minutes. The semen is collected into a collection flask or thermos flask at 37°C through a Buchner funnel fitted with milk filter to filter out the gel plug of the boar ejaculate. Normally a boar will ejaculate 125 to 500 ml of semen. The pre-sperm fraction (about 10–20 ml) contains preputial fluid and urine which should be discarded. Only the sperm rich fraction (100 to 400 ml) is collected. The post-sperm rich part should be discarded which is watery and clear fluid.

Gloved Hand Method

The boar is presented into a semen collection pen and semen is collected using a collection dummy. The boar will mount the collection dummy followed by pelvic thrusting. Semen is collected by applying pressure on the cork screw part of penis by hand is major reflex for boar to ejaculate. Polyvinyl gloves should be used, because latex gloves are toxic for the sperm.

Artificial Vagina Method

Artificial vagina of various designs are developed to collect semen from boars, but are less effective than the manual method. Artificial vagina with a hard rubber cylinder of 12.5 cm length and 4.5 cm diameter with a rubber liner of 40 cm length and 3-3.5 cm diameter may be utilized for semen collection from boars. Small pieces of sponge rubber placed in between hard rubber cylinder and inner rubber lining where hot water at 45 to 50°C is placed, put increased pressure on penis. As the penis passes through the artificial vagina, pressure is applied to the spirally twisted cranial end of the penis by the hand of the operator. The moderate pressure with figures is kept continuous.

Electro Ejaculation

Electro ejaculation method for semen collection in boar is uncommon in boars. Semen is collected under general anaesthesia in lateral recumbency. Rectum is evacuated and a boar-specific rectal probe inserted. The penis is exteriorized and gauze is wrapped around the penile shaft. A pulsating current of progressive amplitude and frequency is applied using an electroejaculator through the rectal probe until ejaculation commences.

Semen Evaluation

The collected ejaculate is brought to laboratory within 15 min for processing and dilution. All the materials and equipment should be sterile to prevent microbial contamination. Semen should maintain at 37°C and materials that come into contact with the neat semen should be at a similar temperature.

Minimum suggested values for neat boar semen obtained via gloved hand technique from adult from crossbred boars on a 4–7 day (Chenoweth and Lorton, 2014)

Semen Variable	Characteristic/Value
Colour	Whitish-grey to white
Ejaculate volume (ml)	216 (range 130–500)
Gross sperm motility	>75%
Sperm concentration (millions/ml)	370 (range 140–735)
Sperm morphology	<20% abnormal
Turbidity	Opalescent

6.3. PRESERVING OF BOAR SEMEN IN LIQUID STATE

Boar spermatozoa are particularly sensitive to cold shock, which induces changes in the plasma membranes of spermatozoa, resulting in a reduction in their biological properties. The effect of cold shock on spermatozoa is dependent on the rate of their cooling, the time and temperature range of storage, as well as the type of applied extender. Extended boar semen is cooled to between 15°- 18°C and is maintained at this temperature for several days until use for AI. Additives which impart antioxidant, sperm membrane stabilization properties and sperm motility enhancer are deliberately used with semen extender for better preserving quality. Extenders should contain ions to maintain the osmotic pressure, glucose as energy source, buffering systems to stabilize the pH of the diluents and EDTA and antibiotics to prevent growth of microbes.

The choice of diluent should depend on its proposed use. If the time of collection and insemination is less than 3 days then short term diluent preferred or if more than 4 days then long term diluent is mostly the choice. But the precaution will be taking care that any loss of sperm viability, fertility and litter size should not affect much every possible effort should be made to ensure the best semen quality.

Two Major Groups of Diluents

- **Diluents for Short-Term Preservation (1-3 Days)**

  Short-term diluents are Beltsville Liquid (BL-1), Beltsville Thawing Solution (BTS), Illinois Variable Temperature (IVT), Kiev, Vital etc.

- **Diluents for Long-Term Preservation (Over 4 Days)**

  The long-term diluents are Acromax, Androhep, Modena, MR-A, Mullberry III, Reading, X-cell, Zorlesco, ZORPVA.

Three different commercial extenders devised for short-term (BTS+) or long term preservation (MR-A and X-Cell), are used to test the storage of semen from four mature, fertile boars at 17°C for 96 hours. The Safe cell plus extender is better than BTS to preserve boar semen in the liquid state at 16-18°C in Large White Yorkshire boars.

6.4. STORAGE OF BOAR SEMEN IN FROZEN STATE

Semen extenders for frozen boar semen are completely different from extenders for liquid semen. There are two major limitations to success in freezing boar semen. First, it is highly sensitive to cold shock, and neither egg yolk nor skimmed milk provides anything like the degree of protection against cold shock that it provides to the spermatozoa of most other species. Second, although glycerol is probably the best cryoprotectant for the boar spermatozoa, its toxic effects are more pronounced in this species than in most others. Boar spermatozoa are very susceptible to cold temperatures and rapid cooling of ejaculates to 15°C or below results in loss of viability. Because unlike bull spermatozoa, composition of the phospholipids in the membrane of boar spermatozoa is different, cholesterol/phospholipid ratio is low and distribution of cholesterol within the membrane is asymmetrical. Usually only 50% or less survives the cryopreservation process. When frozen semen is used, only about five animals can be inseminated per ejaculate.

The current procedure for AI in pigs using frozen thawed semen involves the infusion of a high number of thawed sperm (5000–6000 million) in a large volume of fluid (80–100 ml) into the cervical canal. Despite using such high sperm numbers, there appears to be insufficient colonization of the oviductal sperm reservoirs in the sow's tract, resulting in inadequate numbers of functionally intact sperm in the oviducts to ensure a high fertilization rate. For that reason, AI in pigs with frozen thawed semen often results in conception rates below 20%, far below the rate achieved with fresh semen. Litter size is also lower with frozen than liquid semen insemination.

Different Categories of Freezing Extenders for Boar Semen

Extenders without buffers, such as egg yolk-glucose, egg yolk-lactose and egg yolk-saccharose, ethylenediamine-tetra acetic acid (EDTA), and Mg and Ca salts.

- Extenders with a buffering capacity, such as glycine-phosphate and glucose-phosphate, egg yolk-glucose-citrate, egg yolk-glucose-citrate-EDTA-potassium-unitolurea, Beltsville F3, Beltsville F5, Tes-tris-fructosecitrate-egg yolk, TEST, Tes-NaK-glucose-egg yolk, Tris-fructose-EDTA-egg yolk, and Tris-glucoseEDTA-egg yolk.

6.5. ESTRUS DETECTION AND INSEMINATION OF SOWS

Female Reproductive Tract of Swine

The sow reproductive tract is a long organ compared with other species including human, cow and mare. The ovaries are oval in shape, but in maturity having a mulberry like appearance due to multiple follicles, and/ or corpora lutea. Ovarian follicles are about 7 to 8 mm and corpora lutea about 12 to 15 mm in diameter. Each ovary is surrounded by a thin membrane called infundibulum which act as funnel to collect oocytes and redirect them to oviduct. The oviduct is 15-25 cm long. The uterine body is about 5 cm. long with long tortuous horns. The vagina of the sow is 7 .5 to 11.5 cm. long. It is small in diameter with a thick muscular coat. The vestibule is fairly long. The labia are thick. The minor vestibular glands are variable in number. The clitoris is located about 2 cm. cranial to the ventral commissure. On either side of the cranial portion of the floor of the vestibule there is a culde-sac.

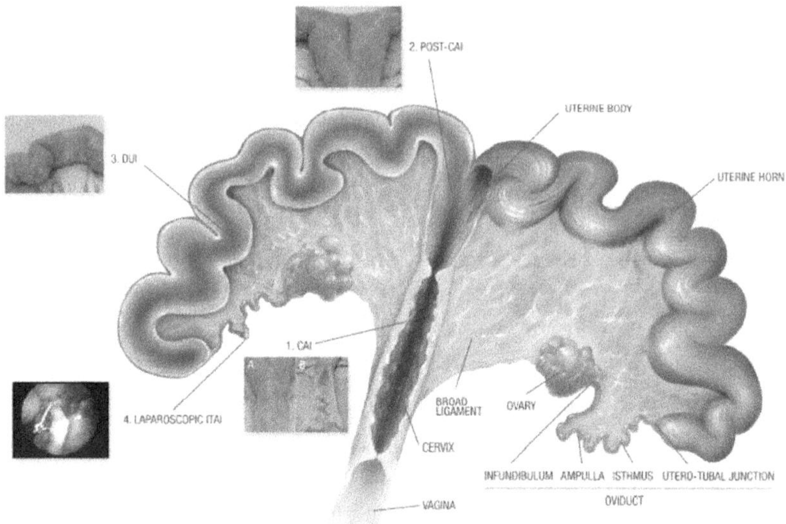

Fig. 2: Anatomy of the sow's uterus and sperm deposition sites during AI. 1) Cervical AI (CAI) (1A and 1B: external and internal view of the cervix, respectively); 2) Post-cervical AI (Post-CAI); 3) Deep intrauterine insemination (DUI); 4) Intraoviductal AI (ITAI) (view of the laparoscopic approach of the oviduct). (Photo source: Soriano-Ubeda *et al.*, 2013).

Detection of Estrus

Three (03) days before estrus vulva becomes progressively swollen and congested which subside gradually 3 days afterwards. Restlessness is an unfailing sign of the approach of heat, and a peculiar repetitive grunt is emitted. The boar is sought, and in his presence the proestrus female noses his testicles and flanks and may mount him but will refuse to be mounted. Estrus can be determined by firmly pressing the loin of the sow with the palms of both hands; the estrus sow will stand motionless with cocked ears.

Timing of Insemination

The optimum time for AI is during a 24 hours period in the middle of the 50–60 hours of estrus. If estrus is first detected during the morning session, the sow should be inseminated for the first time that afternoon, or if estrus is first observed in the afternoon, she should be inseminated the next morning. It is recommended that optimal conception rates and litter sizes are achieved by delaying insemination until 12 to 18 hours after the first detection of estrus, with a second insemination 12 hours later.

6.6. INSEMINATION TECHNIQUE

The procedure of AI should be performed in neat and clean environment and taking into consideration all the sanitary measures including cleaning of the vulva to remove urine and feces and use of a new AI catheter. The catheter is rotated into the cervix until its spiral groove becomes locked into the cervical canal. The cervical lock prevents the catheter from becoming dislodged if the sow moves during insemination and prevents loss of the large volume of fluid that has to be inseminated. However, 90% of the spermatozoa are eliminated within 2 to 3 hours of insemination, and of the approximately 2.3×10^9 spermatozoa that are inseminated, only 1×10^5 become established at the uterotubal junction.

Cervical Insemination

Cervical artificial insemination (CAI) is the most widely used reproductive methodology in the porcine industry around the world. The catheter is easy and simple to apply in the field condition and consist of depositing semen in cervix using catheter. Catheter is with the approximate length of the boar penis and finishing in a corkscrew shaped. The catheter can be categorized into 3 types on the basis of tips such as spiral, foam and multiple ring tip. The catheter should be inserted into the vagina at 45° angle. Slide the catheter until feels a resistance indicating that the catheter at the entrance to the cervix. Then turn the catheter counter clockwise until it locks into the cervix. With a foam or multiple ring tip just slide it into the cervix until it locks. After insemination the catheter is removed clockwise in case of spiral tip and pulled softy outward in case of foam and multiple ring tips.

Post Cervical Insemination

Recently different methods have been directed to reduce the number of sperms inseminated per dose and placing the sperm in different parts of reproductive tract rather than cervix. One of these techniques are post cervical artificial insemination (PCAI) or intrauterine insemination which consist of depositing semen in the uterine body. PCAI is performed with a combined catheter-cannula kit which consist of approximately 72 cm long flexible cannula inserted into a conventional cervical catheter. Unlike the CAI method, the sperm dose should be introduced quickly (few seconds) to spread the dosages through the uterine horns. The inner catheter is removed and then with the cervical catheter still placed in the cervix and shaken in a rotational way, the neck of the womb is massaged for 5 seconds, after which the catheter is removed. This technique needs low number of spermatozoa per dose which increases the number of insemination doses. PCI insemination requires more than 1×10^9 spermatozoa to obtain an acceptable litter size, although pregnancy rates are good even with lower numbers of sperm.

Post Cervical Intrauterine Insemination

This technique was first used by Krueger *et al* (1999). In this technique a surgical deep intrauterine insemination (DUI) is performed using a very low number of sperms deposited closer to the utero-tubular junction. A nonsurgical DUI can be performed by using endoscope without sedation of animal. But the limitation of this technique is cost of endoscope and improper in field condition. A deep uterine catheterization can be performed after insertion of a commercial AI spirette. The DUI catheter is then inserted through the spirette, moved to the cervical canal and propelled forward along with the uterine horn and body. But it has limitation such as high cost, chances of uterine injury and risk of infection.

Intra Oviductal Insemination

This technique permit a drastic reduction in sperm number during insemination and consist of depositing semen directly into the oviduct by laparotomy. This method is called intra tubual artificial insemination (ITAI) and was first used by Polge *et al* (1970).

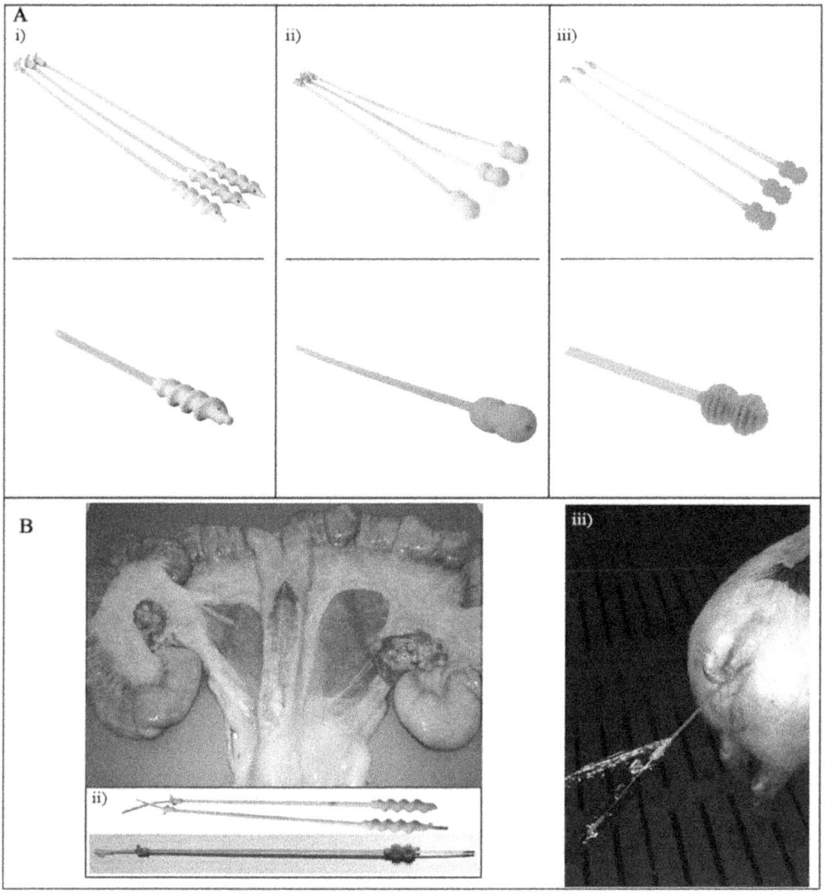

Fig. 3: Catheters used in swine AI.

(A) Different tips used in cervical catheter insemination: i) spiral; ii) foam and iii) multi-ring tip. (B) Post-cervical catheter: i) flexible cannula of approximately 72 cm inserted into a conventional cervical catheter; ii) post-cervical catheter placed in ex vivo uterus, and iii) swine insemination using post-cervical catheter. (Photo source: Soriano-Ubeda *et al.*, 2013).

CHAPTER 7

DISEASES OF SWINE

Dr. Mritunjay Kumar

7.1. COMMON DISEASES IN PIGS

Pig farming contribute substantially in the economy of country and supporting livelihoods for farmers and other personals involved in the pig industry. It converts inedible feeds, forage, meet byproducts, garbage etc into valuable nutritious meat. The piglets and pigs are also susceptible to various diseases like other animals which has detrimental effects on their growth and production. The diseases of piglets and pigs causes severe economic losses due to death, poor growth, and expenditure in the treatment of diseased animals.

7.2. VIRAL DISEASES IN PIGS

7.2.1. African Swine Fever (ASF) /Warthog Disease/African Pig Disease

It is a highly deadly disease of domestic pigs caused African swine fever virus (ASFV) and can cause mortality of up to 100% in affected pigs.

Etiology

ASF is caused by African swine fever virus (ASFV) which is a double stranded DNA virus with an icosahedral symmetry belonged to genus Asfivirus of family Asfarviridae.

DOI: 10.1201/9781003749844-7

Epidemiology

The ASF has been reported in several countries around the world after first report from Kenya in the year 1920.The disease has reported as endemic in Sardinia and outbreaks was reported in 2007 in Georgia, Russia and neighbouring countries. The virus has been reported from India, Bangladesh, Pakistan etc. The disease is limited to all breeds to pigs and all age groups are equally affected. It is distantly related to virus families of lower eucaryotes including the family Faustoviridae and genus Kaumoebavirus.

It is highly contagious and transmitted by direct contact or ingestion of infectious secretions/excretions from infected to susceptible host. Its transmission occurs over long distances through feeding and fomites as ASF is highly resistant in environment. The ASF transmission is also vector-borne which has sylvatic and domestic cycles. In sub-Saharan Africa countries ASFV is maintained as inapparent infections in wildlife hosts such as bush pigs (*Potamochoerus porcus*) and warthogs (*Phacochoerus africanus*) for long period which are infected via tick bites of the argasid sp. (*Ornithodoros* complex).

Pathogenesis

In ASF, clinical signs and pathological changes vary substantially depending on strain virulence and host factors. Acute infections with highly virulent virus strains lead to a clinical course that resembles a viral haemorrhagic fever that is characterized by pronounced depletion of lymphoid tissues, hemorrhagic, or hyperaemic splenomegaly, apoptosis of lymphocyte subsets, and impairment of haemostasis and immune functions. It is generally accepted that most lesions can be attributed to cytokine-mediated interactions triggered by infected and activated monocytes and macrophages, rather than by virus-induced direct cell damage. The petechial and ecchymotic hemorrhages in multiple organs, pulmonary oedema, and disseminated intravascular coagulopathy are generally noticeable.

Clinical Findings

African swine fever occurs in different forms and mortality rate varies between 0 to 100%.

Per-acute form: It is characterized by severe haemorrhagic disease with almost 100% mortality and swine may die without exhibiting any clinical signs.

Acute form: Incubation period varies between 5 to 15 days. High rise of body temperature leads to death of animal between 5 to 10 days after the onset of fever. Such pigs reveal cyanotic area predominantly on the ears, snout, belly, gluteal region, and lower limbs. The animals become recumbent and lie huddled in groups. They revealed frequent shivering, incoordination, unsteady gait, and respiratory distress. Coughing and dyspnoea become predominant under stress. Mucopurulent discharge from nose and eyes, conjunctivitis, vomiting, bleeding from nostrils or rectum and diarrhoea develop subsequently.

Sub-acute form: In this form course of disease is 3-4 weeks and clinically characterized by fluctuating body temperature, cyanotic blotching of skin, weakness, and incoordination.

Chronic form: It occurs mostly in enzootic areas. Emaciation, stunting, pneumonia, ulceration on skin, oedematous swelling of joints, adjacent tissues and mandible and arthritis are common clinical signs.

Post Mortem Examination

The postmortem gross examination shows severe haemorrhagic (dark coloured) splenomegaly with rounded edge. The lymph nodes have multifocal haemorrhage with marbled appearance. The lymphadenopathy is very common in most of the organs. Petechial hemorrhages are often observed in the kidney surface and at sectioning. Multifocal pneumonia is also observed with patches of consolidation and dark color in lung owing to secondary infections due to the state of immune-suppression induce by ASFV.

Diagnosis

- History of presence of ticks in the area and specific clinical signs.
- Blood examination: Leukopenia, lymphopenia and thrombocytopenia.
- Isolation of virus: The virus can be grown on primary culture of pig bone-marrow or peripheral blood leukocytes.

- Serological test: Indirect immune-fluorescence, Immuno-electro-osmophoresis and ELISA are common serological tests recommended for diagnosis.

Treatment: ASF is highly fatal and there is no specific treatment.

Prevention and Control

- Tick control
- The area free of disease should not import pigs or pig products from the areas suffering from the disease.
- The bedding and infected material should be burnt.
- The dead pig must be properly disposed of.
- The premises should be thoroughly cleaned and free of animals for 3-4 months.

7.2.2. Hog Cholera/Classical Swine Fever

It is caused by swine fever virus and characterised by piling of affected animals, discolouration of abdominal skin and nervous signs.

Aetiology

It is caused by swine fever virus (SFV, ss RNA), which belongs to family Togaviridae and antigenically closely related to bovine viral diarrhoea virus. There is only one antigenic type of virus but field strains of SFV can range from highly virulent to almost avirulent. Virus losses infectivity at pH below 4 and above 10 and treatment with ether, chloroform, 3% sodium hydroxide or 5% cresol. Virus remains infectious for months in pork and pork products.

Epidemiology

AFV is worldwide in distribution. Some of the developed countries like USA, UK, Australia, New Zealand, Denmark, Ireland, Norwey, Sweden, Finland, Switzerland, and Canada have eradicated the disease in the last 20th century.

The virus naturally infects only pigs though experimental infection can be produced in laboratory animals as well. Pigs of all age groups are equally susceptible but the severity of illness depends on the virulence of SFV. The highly virulent strains may produce disease in pigs of all age groups while less virulent strains may produce disease in newly borne piglets only. The infected pigs may shed virus before showing clinical signs of disease and during illness, virus is shed in all secretions like saliva, urine and faeces. The pigs that survive acute disease do not excrete virus but chronically ill pigs excrete virus until death.

Transmission

The virus may be transmitted through ingestion or inhalation. Transmission through placental, inanimate objects, porks and common pig louse, flies, and mosquito (mechanical transmission) may also occur. The disease usually occurs in the form of severe outbreaks causing 100% morbidity and almost 100% mortality when virulent strain is involved.

Pathogenesis

The virus enters through oro-nasal route and reach to tonsils (primary replication site) and through lymphatics vessels virus reaches to blood capillaries and initial viraemia develop. The virus is trapped in spleen, peripheral and visceral lymph nodes, bone-marrow and peyer's patches. Then there is secondary viraemia and virus invades parenchymatous organs. The predilection site for ASF virus is reticular, endothelial, and epithelial cells. Most of the lesions are produced by hydropic degeneration and proliferation of vascular endothelium that occurs due to occlusion of blood vessels. As a result of which congestion, haemorrhages and infarction in arterioles, venules and capillaries and thrombosis in small and medium sized arteries occur.

In pregnant sow, transplacental spread occurs as virus multiply at one or two site in placenta and may spread subsequently from foetus to foetus. Usually low to moderate virulent strains of SFV are involved in congenital infections. In early foetal stage, more damage to the developing foetus and prenatal mortality occurs. The virus has affinity for cells of immune system which can lead to alterations in immune reactivity.

Clinical Findings

Acute infection: Incubation period in acute infection varies between 2 to 6 days. The young pigs may die suddenly without showing symptoms and other animal may reveal rise in body temperature (106-107^{0}F), reduced appetite, dullness, reluctant to move and leukopenia. conjunctivitis, marked ocular discharge, mucoid to purulent nasal discharge and constipation followed by diarrhoea may be seen. Affected pig shiver and pile upon each other. In some cases, convulsions while in terminal stage, typical weaving, staggering gait, and posterior paresis are seen. A purple discolouration extending over abdomen, snout, ears, and medial sides of legs may also be observed. Circling, inco-ordination, muscle tremor and small area of necrosis on the edge of ears, on tail and lips of vulva. About 90-95% of affected pigs die between 10 to 20 days post infection.

Persistent Infection: When the pig survives for 30 days, it is usually referred as persistent infection and, in such cases, there is chronic infection. The affected pigs reveal anorexia, fever, diarrhoea and alopecia followed by dermatitis, blotching of ears and deep purple coloration of abdominal skin. It may lead to development of runt pigs due to severe growth retardation and such pigs have relatively large head and small trunk. The pigs stand with arched back and hind legs placed under the body.

Congenital Infection: Most piglets infected in utero and borne alive but they die shortly after birth. In such piglets, haemorrhages over skin are commonly seen. In congenital infections abortions, faetal mummification, stillbirths and neonatal deaths are usually noticed. The born piglet usually remains weak and reveal tremors. Other piglets may look healthy but carry persistent SFV infection and may show signs of disease. Such animals die after 2-11 months of age.

Necropsy Findings: Multiple haemorrhage of varied size is seen under the capsule of kidney, about ileo-caecal valve, cortical sinus of lymph nodes and in bladder and larynx. Infarction of spleen is pathognomonic of disease and occurs as dark blebs of various sizes, slightly raised above surrounding surface. Congestion of liver and bone marrow, and circular, raised button ulcers in colonic mucosae are indication of the disease.

Microscopic Examination: Perivascular cuffing in brain, non-suppurative encephalitis, atrophy of thymus and severe depletion of lymphocytes in tonsil, lymph nodes and spleen, and degeneration of endothelial cells. Chronic form of disease is characterized by necrotic ulceration of mucosa of large intestine and transverse calcification of the distal portion of rib.

Diagnosis: History and clinical signs: History of outbreak and clinical signs of the disease.

Blood examination: Leukopenia (4000 to 9000/μl) with thrombocytopenia is consistent findings. However, in terminal stage leukocytosis occurs due to secondary bacterial invasion.

Isolation of Virus: For the growth of SFV, porcine cell culture is most frequently and PK-21 cells are also used for this purpose.

Serological Test: The immune response in host is evoked in 2-3 weeks, so for detecting antibodies blood sample should be collected 3 weeks or later. The serological test like VNT, FAT, AGD, passive HA, CFT and indirect ELISA are useful. Monoclonal antibodies using ELISA can be useful in differentiating the SFV and BVD virus. FAT can be employed on frozen section of tissue as well as on impression smears prepared from fresh tissues.

Differential Diagnosis: ASF should be differentiated from salmonellosis, acute erysipelas (ecchymotic sub serous haemorrhage and diamond shaped skin lesions) and African Swine Fever.

Treatment

No specific treatment only symptomatic.

Acute cases may be given hyper immune serum as early as possible @ 50-150ml and can be repeated after 24 hr.

Prevention and Control

Isolation and treatment of affected animal in endemic area. Slaughtering of all the infected or in-contact pigs, and such pigs after slaughter are properly disposed of by burning. Maternal antibodies provide immunity to young one up to 5 weeks of age as it crosses the placental barrier. ‘ The pen of infected pigs must be properly washed with 5% cresylic acid solution or other suitable disinfectant.

Pork or pork product prepared from enzootic areas should not be used. Such products must be burnt. Crystal violet vaccine (inactivated): provides immunity in 2 weeks of vaccination but duration of immunity is very short and pig requires repeated vaccination.

Modified live virus vaccine: Rapid long-lasting immunity but capable of causing infection during vaccination.

Tissue Culture Vaccine: Piglet born from immunized sow between age group of 8-10 weeks and piglet born from un-immunised sow at 2 weeks of age are immunized. It provides immunity for 3 years. In these piglets booster dose is given at 6-7 months later. *Swine fever vaccine (Intervet/IVRI)* first dose at 8-10 weeks of age and booster dose is given annually @ 1 ml intra-muscularly.

7.2.3. Pseudorabies (Mad Itch, Aujeskey's Disease)

It is an infectious disease mainly of pigs and occasionally of cattle, sheep, dogs and cats, caused by herpes virus. In young pigs, it is characterised by acute fatal encephalitis and in other species it is characterised by both encephalitis and marked local pruritis.

Etiology

It belongs to family herpetoviridae and genus herpes virus. It is a DNA virus with cubical symmetry and envelop. It replicates inside the nucleus and produces intra-nuclear inclusion bodies. It can grow in developing chicken embryo and can be passaged in rabbits and unweaned hamster.

Epidemiology

Incidence: Pseudo-rabies occurs in many countries of the world. Pigs are commonly affected and rarely cattle, sheep, dogs and cats are affected. Human beings can be affected but it is not fatal. In cattle it is usually fatal.

Transmission: Not clearly understood. It has been shown to spread from pigs to pigs by contact. But it will not spread by contact in other domestic animals.

Shopes hypothesis: According to this hypothesis the cattle suffered from pseudorabies and they die. The rat will ingest small number of carcasses and then they migrate to the piggery unit and die. The pigs ingest the rats and they will come down with this disease. The portal of entry is through the abraded skin or the intact nasal mucous membrane.

Pathogenesis

The virus gets into the circulation and there will be viraemia for a short period. Then the virus localises in the respiratory tract and replicates. Through the olfactory, glosso-pharyngeal, or trigeminal nerve gets into the brain and causes encephalitis. When the virus gets through the abraded skin it gets into the peripheral nerve and causes irritation which is responsible for the pruritis.

Clinical Findings

In young piglet i.e. less than 1 month of age the disease is very severe. In the beginning high fever of 107°F and the incoordination of hind limbs and recumbency are seen. Tremors, paddling movements, lateral deviation of head, frothy salivation, nystagmus, and convulsions occur. Within 12 hours death can occur. In some pig blindness due to an extensive retinal degeneration can occur.

The disease is very mild in adult swine. The signs noticed are anorexia, dullness, agalactia and constipation. Mummified foetus, stillbirths, late abortions can also be seen. Pruritis is absent in young piglets.

Necropsy Findings: Considerable damage to local areas of skin and extensive subcutaneous oedema, pneumonic changes in lung, haemorrhages in the endocardium and the fluid accumulation in pericardial sac.

Histo-pathologically encephalitic changes are seen (perivascular cuffing, degenerative changes of neuron). Intra-nuclear inclusion bodies are seen in the neurons of the cerebral cortex of pigs.

Diagnosis

The pseudorabies may be diagnosed based on history and occurrence of intense pruritis and encephalitis in piglets.

Histopathologically intra-nuclear inclusion bodies present. Sample to be submitted are local tissue of pruritis, part of local nerve trunk and pieces of cerebellum.

Animal Inoculation: The suspected material is injected into the rats or mice and they come down with the disease.

Serological Test: FAT, SNT, VNT, ELISA are useful diagnostic aid.

Treatment: Treatment for pseudorabies is not available.

Control

- Destruction of rats.
- Rearing of cattle and pigs separately.
- Vaccination: Killed and attenuated vaccines have been used. The vaccine is administered intra-muscular and peripherally. No vaccine is available in our country.

7.2.4. Transmissible Gastroenteritis (TGE)

It is highly contagious disease of pigs caused by transmissible gastroenteritis virus (TGEV) and is characterised by diarrhoea, vomition and high mortality rate.

Etiology

It is caused by TGEV which belongs to coronaviridae. It is RNA virus and measures 80-100 nm, helical symmetry and enveloped. The envelop is having petal shaped projections. It is cultured in porcine kidney cell culture. Cytopathic effect is characterised by rounding off cells and detachment of cells. It is susceptible to beta propiolactone and saponin.

Epidemiology

Transmissible gastroenteritis (TGE) occurs throughout the world. Very important disease of swine industry causing severe economic loss in the form of mortality rates (100%) of piglets under 2 weeks of age. The virus is photosensitive and hence the disease is not common in summer but common in winter. It is transmitted by ingestion and inhalation.

Pathogenesis

The virus upon ingestion and inhalation goes to small intestine and it is having very affinity for tall columnar epithelial cells of the villi of the small intestine. The virus replicates in the cells and the cells are destroyed. Rate of destruction is more than rate of replacement; hence contraction of the villi occurs and is covered by low cuboidal epithelial cells. This results in the shortening of the villi.

The above changes results in reduction in production of certain enzymes required for digestion. Hence, lactose is not hydrolysed and not absorbed. Presence of lactose increases the osmotic pressure which results in withdrawal of lots of fluids leading to diarrhoea-dehydration-haemo-concentration. Here also metabolic acidosis and hypokalemia occurs which may be responsible for death of piglet.

Clinical Findings

The incubation period of TGE is 18 hours to 3 days. The vomition followed by diarrhoea generally occurs. Faeces will be watery yellow in colour and foul smelling and it contains small curds of undigested milk. Signs of dehydration are also seen. The course of the disease is 2-7 days.

Necropsy Findings

The stomach and small intestine are congested and contains curdled milk. The intestinal wall is thin and almost transparent. In normal pigs chyle which is white or milky fluid and contain lymph and emulsified fat present in the mesenteric lymphatics. But in TGE it is absent. Examination of villi with a hand lens showed villi: crypt ratio 1:1 (normally it is 7:1).

Diagnosis

- By signs and lesions.
- Culturing: Faecal swab, piece of small intestine.
- Serological test: VNT, FAT, IPT, HIT, Competitive ELISA.
- Presence of intestinal lactose enzyme.

Differential Diagnosis

- Colibacillosis: No vomition.
- Clostridium perfringens type C: Haemorrhages in the intestine, blood in the faeces.
- Swine dysentery: Adult swine are commonly affected.

Treatment

- To dehydration can be manage by administration of lactated ringer solution i.p., glucose like electrolytes.
- To overcome hypokalaemia-potassium to be supplemented.
- The metabolic acidosis can be managed by administration of sodium bicarbonate.
- The vomition can be manage using perinoram/stemetil and diarrhoea by using astringent mixture.

Control

Good hygiene measure.

Vaccination: Moderately virulent TGEV is administered orally to pregnant sows is highly beneficial. Hence, the IgA producing B-lymphocytes present in the intestine was exposed to the virus and these lymphocytes gets into circulation and gets localised into the mammary gland. After farrowing these lymphocytes become plasma cells and start secreting IgA. The piglets when suckle colostrum, they ingest contain good quantity of IgA which goes to the intestines and coats the mucosa. The IgA is resistant to acidic pH of stomach and hence good amount of IgA present in the intestine. When such piglet ingests virulent TGEV it can bind to the epithelial cells and cannot replicate and thus cannot produce the disease. Hence, this is most used method for the prevention of TGE.

7.2.4. Porcine Reproductive And Respiratory Syndrome (PRRS)

PRRS is caused by arterivirus and characterized by low farrowing rate, marked increase in abortions, mummified and weak live born piglets.

Etiology

It is caused by an RNA, arterivirus of order nidovirales and family arteriviridae. There are two antigenically and genetically distinguishable strains, genotype 1, with the prototype lelystand virus representing the viruses predominant in Europe and genotype 2, represented by VR 2332, commonly prevalent in America and a variant of genotype 2 in Asia. Pig and other members of family suidae is only susceptible.

Epidemiology

It was first recognized in North America in 1980 and now present throughout the world. PRRS is not prevalent in Australia, New Zealand, Finland, Norway, Sweden, and Switzerland. The PRRS virus is moderate resilient to environmental degradation. The PRRS virus is easily deactivated by phenol, formaldehyde, and most common disinfectants. It occurs in all age groups of pig and wild pigs.

Pathogenesis

The virus of PRRS compromising the cellular immune response, damage mucosal surface and has an affinity for macrophages. The replication of virus mainly occurs in macrophages of either lymphoid tissues and/or lungs in the acute phase of infection. The PRRS virus persists in tonsil and lung macrophages. The virus antigen has been found in the inhabitant macrophages of a variety of tissues, as well as in other cells, including muscle tissues.

Clinical signs

The incubation period of PRRS is 14 days. Inapparent clinical signs has observed in many infected pigs in herds. The appearance of clinical signs among pigs varies and are influence by many factors like virulence of virus. Where signs are apparent, they vary and are influenced by several factors such as age of pigs, herd size, virulence of the virus and managemental practices. Reproductive impairments are commonly reported from sows or gilts and from some boars. The respiratory syndrome is more common in young growing pigs but also occurs in immature finishing pigs and breeding stock.

Breeding Age Gilts, Sows, and Boars

Clinical signs includes anorexia, fever, lethargy, and perhaps respiratory distress or vomiting. In few outbreaks, mild cyanosis of the ears, abdomen and vulva has been reported. The reproductive problems are chief clinicals signs characterized by decrease in conception rate or farrow rate in dam, increase in premature farrowing, late term abortions, stillborn or weak piglets, high preweaning mortality and mummified fetuses. Nursing pigs may have dyspnea ("thumping") while boar is accompanied by decrease in semen quality. The infected sows or boar may act as source of infection to other animals in the herd.

Young, Growing and Finishing Pigs

The young pigs generally show respiratory problem specially in 4 to 10 weeks age group. The piglets show sneezing, fever, depression, lethargy, stunting, pneumonia and followed by expiratory dyspnea and stunting.

Diagnosis

The clinical signs and history are useful in diagnosis of PRRS as there is no pathognomonic lesions. The diagnosis may be done by virus isolation (VI), detection of PRRS antigen by fluorescent antibody tests (FAT) or immunohistochemistry (IHC), or detection of PRRS virus genome by polymerase chain reaction (PCR). The antibody can be detected by ELISA kits which is commercially available for PRRSV antibody detection but serological test can give false positive and false negatives on individual animal but on a group basis it is reliable indicator. RT-PCR has been used to detect viremic animals or carriers using serum, tonsil biopsies or oro-pharyngeal scrapings.

Treatment and Control

The specific treatment is not available against PRRS. Broad-spectrum antibiotics and anti-inflammatory products are generally used during acute disease. In confirmed cases of PRRS, common strategy used is to either eliminate the disease or to control (or "live with") PRRS. The control measures include the management of arriving replacement gilts, the use of vaccines and execution of biosecurity protocols authorized to reduce the risk of PRRSV spread within and between herds. The whole herd depopulation/repopulation, test and removal, and herd closure are common methods of eliminating virus from endemically infected herds.

Vaccine

Prime Pac™ PRRS RR vaccine (Intervet Inc.) Sows and gilts: The single injection of 1 ml dose intramuscularly (IM) 8 weeks prior to breeding is recommended. Piglets: Inject a single 1 ml dose IM at 3 weeks of age or older. The revaccination frequency depends on age of sows. Duration of immunity against reproductive and respiratory form of the disease is at least 20 weeks and 23 weeks, respectively.

7.2.5. Swine Influenza (SWINE FLU)

Swine influenza is a highly contagious respiratory disease caused by Influenza A virus and characterized by anorexia, fever, coughing, sneezing, lethargy, and abortion.

Etiology

Swine influenza A virus (IAV) belongs to the Orthomyxoviridae family, is an enveloped, negative-stranded RNA virus with a segmented genome. It has antigenic proteins namely hemagglutinin and neuraminidase which is responsible for viral attachment to cells, virions release form infected cells and determine the virus sub type. The sub types of influenza which causes infections in pigs are H1N1, H1N2 and H3N3 with multiple strains with each subtype. The influenza B and C viruses have been isolated but not responsible for any classical disease. The mixing of carrier animals with susceptible pigs is a major predisposing disease. The co-infection of PRRS and Glasser disease, Actinobacillus pleuropneumoniae and M. hyopneumoniae are commonly associated with IAV. The virus can be readily neutralised by disinfectant.

Epidemiology

Swine influenza is endemic and worldwide in distribution. The infection mainly transmitted by aerosolization (limited distance), direct contact, and fomite transfer. Generally, one or two individuals in a herd affected initially followed by infection spread rapidly within the herd. The evidence of IAV infection has been reported in Europe, North America, South and Central America, and Asia. Carrier pigs, introduces IAV into previously uninfected herds and countries.

Pathogenesis

Swine influenza infection varied from sub-acute to acute form. The virus through inhalation enters and multiplies in bronchial epithelium within 16 hours of infection. It causes focal necrosis of the bronchial epithelium, focal atelectasis, and gross hyperemia of the lungs. The individual lobules of apical and intermediate lobes within 24 hours of infections shows bronchial exudates and widespread atelectasis, as plum-colored lesions. The lung lesions after infections continue to develop until 72 hours after that the virus becomes hard to demonstrate. The IAV very rarely causes systemic infections and losses due to reproduction may be associated with high fever.

Clinical Signs

The disease is characterized by sudden onset, rapid spread in entire herd within 1-3 days and sudden recovery. The clinical signs are flue like high fever (upto 108°F), depression, anorexia, coughing, nasal and ocular discharge and mortality rate varies between 1% to 4%. The IAV may continue to circulate in herd as reinfection in recovered pigs which may occur from same or distinct strain. The economic losses in IVA may occur due to stunted growth, death of piglets and reproduction problem like abortion, infertility, embryonic re-absorptions, and late pregnancy.

PM Findings

The thoracic cavity is normally affected in uncomplicated cases and lung are clearly demarcated, collapsed, and purplish red throughout the lung (patchy pneumonia) on gross examination. The unaffected lung areas are pale and emphysematous. The airways contain mucopurulent discharge while lungs may have severe pulmonary edema of interlobular septae, or a serous or sero-fibrinous pleuritis.

Diagnosis

Based on clinical signs and history

Virus isolation and RT PCR are commonly used for confirmatory diagnosis.

Treatment and Control

There is no effective treatment although antimicrobial along with supportive treatment may be given. Good management practices such as all in/all out procedures, limiting movement of pigs and sow, stress free environment reduces the transmission of disease. The vaccination of person involve in interaction with pigs is useful in prevention of bidirectional transmission of swine influenza between pigs and people.

7.2.6. Porcine Parvovirus (Ungulate Proto-parvovirus 1)

Porcine parvovirus (PPV) is most common cause of reproductive failure in swine and characterized by acronyms (SMEDI) stillbirths, mummification, embryonic death and infertility when sows and gilts are exposed to virus between 6 to 70 days of gestation.

Etiology

Porcine parvo virus is non-enveloped DNA virus of parvoviridae family.

Epidemiology

The pig is the only host to porcine parvovirus and prevalent worldwide. The virus is stable in regards to temperature and chemicals and can be transmitted through fomites such as clothes, boots, equipment etc. It spread oro-nasally through faeces and other secretions from infected pigs.

Pathogenesis

The porcine parvo virus after entry through oro-nasal route replicates initially in tonsils after which it reaches the lymphatic system leading to cell free viraemia. The virus requires host DNA polymerase to replicate thus can only produce virus particle in S-phase. The virus has affinity to invade rapid dividing cells.

Clinical Signs

Acute infection of both post-natal and pre-natal dam is sub clinical and the pig will have transient leukopenia within 10 days of infections. Infertility, abortion, still birth, neonatal death, prolonged gestation, and less viability of neonates are common clinical signs. Most of the infected foetus will have crown rump length of 17 cm or less because those infected after day 70 are able to overcome the viral assault and survive.

Diagnosis

Mummified foetus that has crown- rump length is ≤ 17 cm strongly showed PPV infection. The viral antigen identification by immunofluorescent (IF) microscopy from sections of fetal tissues is advocate for confirmatory diagnosis. Serologic testing like ELISA is advised when antigen from mummified foetus is not available.

Differential Diagnosis

PPV virus should be differentiated from PRRSV, Brucellosis, Leptospirosis, PRV and Toxoplasmosis.

Treatment and Control

There is no treatment for reproductive disorder in this disease.

The best way to prevent the PPV is either naturally infect the gilt or vaccinate with a modified live vaccine. They are vaccinated several weeks before conception. Vaccination of boar is also advised.

7.2.7. Porcine Circovirus

Porcine circovirus diseases have traditionally been defined as those associated with porcine circovirus 2 (PCV-2) infection.

Etiology

PCV is non-enveloped single-stranded, circular DNA genome virus which is non-cytopathogenic, picornavirus-like contaminant in the porcine kidney cell line PK-15. Out of 3 porcine circoviruses, PCV-2 and probably PCV-3 are pathogenic. Phylogenetic studies revealed at least four genotypes of PCV-2 (PCV-2a, b, c, and d).

Epidemiology

It is ubiquitous in countries with and without PCVDs. PCV-2 and PCV-3 infections reported in wild boars besides PCV-2-SD also present in wild boars. PCVDs associated with PCV-2 have been reported worldwide.

The PCV is transmitted by direct contact with diseased pigs. PCV-2 is present in all secretion and excretion from the body of infected pigs. The virus is also present in semen, but the practical importance of this is insignificant.

Pathogenesis

The viral infection of PCV-2 in pigs damage to the immune system causes acquired immunodeficiency. The virus causes depletion of lymphocyte of lymphoid tissues, changes in blood monocytes populations, and changed cytokine expression patterns have all been demonstrated. The accumulation viral particles in the macrophages and dendritic cells of diseased pigs causes appearance of large amount of PCV-2 antigen. The epithelial and endothelial cells appear to be the main target for PCV-2 replication.

Clinical Signs

The most noticeable sign in the acute phase is the existence of irregular, red-to-purple macules, and papules on the skin of the hindlimbs and perineal area. In severely affected animal distributions may be generalized. The lesions become covered by dark crusts and fade gradually with or without scar. The affected pigs have mild fever, anorexia, depression, prostration, stiff gait and/or reluctance to move.

PM Findings

The liver may appear icteric and/or atrophic in few affected pigs with prominent interlobular connective tissue. The enlarged kidneys show scattered to diffuse white foci on the cortical surface. Microscopically lesions vary from single cell necrosis (apoptosis) with mild lymphocytic infiltration of portal zones to extensive lympho-histiocytic periportal hepatitis with diffuse necrosis of hepatocytes.

Diagnosis

The clinical signs of wasting or ill thrift, still birth and non-viable neo-natal piglets and gross lesions of congestion of liver, hypertrophy of heart and myocardial discolouration. Microscopically, lymphoid lesions, extensive fibrosis and necrotizing myocarditis reported. ELISA, indirect fluorescent test or immune-peroxidase staining of infected cell culture are serological test used for the detection of antibodies against PCV-2. The PCV may be isolated on several porcine cell line by using tissue homogenate, serum and bronchial lavage. Serum or tissues of infected pigs can be used for detection of Viral DNA.

Treatment

Treatment using wide range of antibiotics is not effective in PCV-2 infection.

Prevention and Control

The use of antibiotics to combat secondary bacterial infection, strict biosecurity and sanitary measures and control of concurrent viral infection especially of PRRS are effective. Currently, PCV-2 vaccines are used to control the PCV-2-SD and PCV-2-SI infection in pigs. The inactivated PCV-2 isolate is commercially available for use in piglets, sows, and gilts. Piglets are vaccinated at the age of 2 to 3 weeks.

7.2.8. Foot and Mouth Disease (FMD) (Aphthous fever, Contagious aptha, Panzootic aptha)

It is a highly contagious animal disease of all cloven-footed animals characterised by fever and vesicular eruptions in buccal cavity, teats and feet.

Aetiology

The disease is caused by aphtho virus belonging to family Picornaviridae. It is positive sense single stranded RNA virus and is the smallest known virus of animal origin. There are seven major serotypes *viz* O, A, C, Asia 1 and SAT 1, SAT 2 and SAT 3. These serotypes were originated from different places, viz O and A in France, C in Germany, Asia 1 in Pakistan and SAT 1, SAT 2 and SAT 3 in South African Territories. All the strains differ in antigenicity and virulence. Many subtypes have also originated, *viz* O1, A5, A10, A22, A79, A85 and A87. In South East Asia, O is most prevalent and C is least and type A22 specifically has gained upper hand in some areas of India.

The virus may remain viable for about 1 year in infected premises, 10-12 weeks in clothing and feeds, 4-5 months in hay and bran, 3-4 months in lymph node and bone marrow and one month on hairs and in infected frozen bull semen. It is stable at pH 7.4 to 7.6 but is inactivated at pH below 6. Virus in milk is killed after boiling for 15 minutes. It is resistant to alcohol, ether and chloroform. It can be destroyed by 1-2% sodium hydroxide or formalin and 4% sodium carbonate solution in few minutes.

Epidemiology

FMD is probably the world most important animal disease even though the mortality is low. Loss of productivity, expense of eradication and interference with movement of cattle and meat between the countries are the most important economic effects. The prolonged convalescence impairing both milk and meat and its sequelae like mastitis, abortion, sloughing of hooves and panting and diabetes mellitus are the serious effects of FMD. It is a mild zoonosis too and pigs are considered as amplifier host of FMD.

The disease is widely prevalent in South America, Africa, Central Europe, and Asia. It stands now eradicated from Australia and New Zealand have not experienced the disease since more than a century. The disease is generally found in the form of outbreaks and spreads fast from herd to herd. In India, the average annual incidence of FMD has recently been reported about 23%, being highest in western part (35%) and lowest in southern part (11%). The morbidity rate during outbreak approaches 100% but case fatality rate is very low, about 2% in adults and 20% in young calves which may develop a malignant form of disease. The disease occurs in all seasons and the seasonal peak is related to mass movement and congregation of animals caused by migration to new pasture and cattle or pig fairs.

The host range is wide as cattle, buffalo, sheep, goat, pig and yak are the common victims of natural outbreak of disease. Pigs and sheep may act as important carriers of disease. Immature animals are more susceptible.

Infected animals, carriers, infected materials, meat, semen and other animal products are the source of infection. The virus is present in milk, saliva, urine, faeces and semen etc in sufficient quantity. Animal to animal transfer occurs by direct contact. Ingestion and inhalation are the main portal of entry. Air borne infection through wind currents particularly in humid atmosphere may carry the virus over the long distance as far as 250 km.

In recovered animals, naso-pharynx and mammary tissues are the main sites of virus persistence. Wild fauna may also serve as reservoirs and human beings may transport it from herd to herd. Movement of animals, transportation of virus on inanimate objects (meat, milk, and semen), birds carriage, and contaminated vaccine have been attributed as the cause of transport of infection between herds, areas and countries or continents (Transboundary animal disease).

Pathogenesis

Irrespective of route of entry, virus gets into the blood and there will be viraemia for a short period. Then the virus localises in the epithelium of the oral cavity and inter-digital space and in some animals also on the teat and udder. The virus causes hypertrophy and hyperplasia of epithelium, leading to formation of vesicles. The small vesicles fuse (coalesce) together to form bulla. The vesicle/ bulla is thin walled and will be filled with clear straw coloured fluid. Because of prehension, mastication and walking, the vesicle will rupture leading to formation of ulcers.

Clinical Findings

Incubation period is 1-21 days but commonly it is 3-8 days. In pigs, clinical signs of high fever (41.5°C), lameness, and blanching around coronary band present. The affected pigs show little interest in feed, huddling among other pigs and lethargy. The vesicles develop on the coronary band and the heel of the foot including accessory digits, snout, tongue, and mandible. Besides, pigs reared on rough surfaces, vesicles may be present on the hocks and knees of pigs and based on the severity of vesicles, the horn of the foot slough off and causes chronic lameness. Piglets, below 14 days of age, may die suddenly without any clinical signs due to virus-induced myocarditis.

Necropsy Findings

Vesicles, bullae and ulcers in the oral cavity and tongue are commonly seen. These lesions also can be seen in pharynx, oesophagus, fore stomach, and intestines and sometimes also from the trachea to bronchi. In young piglet myocarditis is usually present.

Diagnosis

Clinical signs and lesions

Presence of vesicles on the dorsum of tongue and lesions in the inter-digital space legs.

Isolation of the virus

The vesicular fluid, tongue epithelium, saliva, faeces, milk, oesophago-pharyngeal fluid should be collected and cultured in BHK-cell, tongue epithelium and in developing chicken embryos, Guinea pigs and unweaned mice.

Serological Test

Neutralization test, CFT, FAT, IPT, ELISA and the virus infection associated (VIA) antigen test have been used for the detection of FMD.

Differential Diagnosis

The disease to be considered is Vesicular stomatitis, Vesicular exanthema, Blue tongue, Bovine Viral Diarrhoea and Rinderpest.

The details of animal inoculation test are as follows:

Sp. of animal	FMD	Vesicular Stomatitis	Vasicular Exanthema	Blue tongue
Cattle	+	+	-	+
Pig	+	+	+	-
Sheep & Goat	+	+	-	+
Horse	-	+	-	-
Adult chicken	+	+	-	-

Treatment

There is no effective treatment for this virus. However supportive therapy is given.

- To overcome the secondary bacterial invaders, administer a course of broad-spectrum antibiotics like Dicrysticin, Vetopen etc i/m for adults for 5 days.
- To reduce the fever and pain administer antipyretics and analgesics like analgin, meloxicam, ketoprofen etc.
- Wash the mouth with any one of the following solutions. Potassium permanganate (PP) solution (0.01%), 2% alum, 2% copper sulphate, 0.5-1% NaOH, 1-2% Sodium carbonate. Then apply boroglycerine or borax glycerine.
- Clean the foot lesion with PP solution and then apply any one of the preparations like Lorexane cream (Benzzedine hexachloride)/Himax/ Neosporin twice daily in lesions.

Prevention and Control

The control of FMD is very difficult because FMD virus readily undergoes mutations and highly potent vaccine inducing immunity for long period is not available.

- *Testing and slaughtering*: For the control and eradication of this disease in some countries testing and slaughtering of reactors and in contact animals has been followed and has given good results.
- In developing countries like India generally isolation of affected animals, treatment of animals and general hygiene measures are followed.
- Vaccination of healthy animals is measures followed for the control of FMD. Ring vaccination to contain outbreak is followed in some countries, besides border line vaccination.

Vaccine

The vaccines used in our country are all killed vaccines and aluminium hydroxide adsorbed vaccines. They are all polyvalent vaccine and produce immunity for 6-8 months; hence they should be repeated in a year i.e. twice a year. Since in most of the places FMD outbreak occur in February or March or afterwards and the animals are commonly vaccinated in November/December/January.

The latest vaccine produced is by using genetic engineering technology or gene splicing or DNA recombinant technology in USA. Out of these, viral protein-1 (VP-1) is the most important one which is involved in the immune response against this disease. Then the gene responsible for VP1 is identified and separated and incorporated into E.coli to produce VP1. The protein is isolated and purified and used as vaccine (Dose 250μg/ animal). Raksha Ovac (Indian Immunological): It contains inactivated tissue culture FMD virus. Antigens O, A, C and Asia I adjuvanted with mineral oil can be given @ 2ml in pig at 2 months of age then after every 6 months of intervals.

7.3. BACTERIAL DISEASES IN PIGS

7.3.1. Swine Erysipelas

Erysipelas in swine is caused by *Erysipelothrix rhusiopathiae* and characterized by erythema, diamond shaped lesions, septicaemia, arthritis and endocarditis.

Etiology

It is caused by *Erysipelothrix rhusiopathiae* in growing and adult swine.

Epidemiology

The bacteria commonly reside in the tonsillar tissue and shed the organism in their faeces or oro-nasal secretions which acts as important source of infection for other swine. The infection may occur through contaminated faeces and water, and aberrated skin. In farm with endemic infection of erysipelas, young piglets get natural infection but clinical signs appears only when they become adult due to maternal derived passive immunity. Healthy pigs may also act as carrier.

Pathogenesis

The erysipelas enter into the body through palatine tonsil or lymphoid tissue present in gut, and also via wounds in skin by direct contact or arthropod bites. The pathogenesis of the disease is not entirely understood, but virulence factors are only partially characterized. Besides, other important factors may contribute like neuraminidase, capsular polysaccharides, and surface proteins.

Clinical Signs

The swine erysipelas outbreaks may occur as acute or chronic and clinically inapparent infections. Acute infection characterised by sudden death, fever, painful joint, inappetence and skin lesions vary from diamond skin (rhomboid urticaria) lesions to generalized cyanosis.

A chronic erysipelas is vegetative valvular endocarditis where valvular lesions show few clinical signs. The signs of respiratory distress, lethargy, and cyanosis, and possibly suddenly death to the infection reported on physical exertion.

The septicemic form occurs most frequently in growing and finishing pigs. The pigs may walk stiffly on their toes and while standing shift weight from limb to limb. Generally, death may occur up to 6 days after the appearance of first clinical signs. Mortality rate varies between 0-100%. Skin lesions may vary from erythema and purple discolouration of the ears, snout, and abdomen to diamond shaped skin lesions over the body. The pink or purple area of variable size become raised and firm within 2-3 days of illness and later either disappear or progress to more chronic type of lesions referred to as diamond skin disease.

Pregnant show may abort due to fever and lactating sows may show agalactia. In chronic form pigs may develop chronic arthritis, vegetative valvular endocarditis, or both. Valvular endocarditis commonly manifested by death is frequent in mature or young adult pigs.

Post Mortem Findings

In acute death due to erysipelas pigs does not have any discrete lesions. The cyanosis diamond-shaped skin lesions may be present. Congested lymph nodes are often oedematous along with sub-scapular haemorrhage.

Diagnosis

Diagnosis of erysipelas is frequently based on clinical signs of fever, lameness and rhomboid urticaria or diamond skin lesions is almost diagnostic. post mortem examination and

E. rhusiopathiae isolation from blood of diseased animals, particularly after enrichment is helpful in diagnosis of acute cases. Besides, molecular methods for erysipelas DNA detection in affected tissues or blood (i.e., PCR assays) is useful for confirmatory diagnosis. Immunohistochemical methods are useful in diagnosis of cases where pigs have been treated with antimicrobials before sample collection.

Treatment

E. rhusiopathiae should be treated with penicillin at 12-hour intervals for a minimum of 3-5 days as best choice. Ampicillin and ceftiofur is also good in acute cases. In case of outbreaks oxytetracycline may be given in the feed or water. The NSAIDs such as flunixin meglumine is useful to overcome fever. Erysipelas antiserum is reported as an effective therapy in acute cases although chronic cases are not responding to treatment.

Prevention and Control

Vaccination is very effective in controlling the outbreak of swine erysipelas in herd. Live attenuated vaccines and bacterins are commonly used to provide extended duration of immunity. In endemic area susceptible pigs should be vaccinated before, after and at the time of weaning. Breeding stock usually vaccinated twice a year. Newly added pigs in farm should be vaccinated with a booster 3–5 weeks later.

7.3.2. Glasser Disease

Glasser disease in pig is caused by *Glaesserella parasuis* and characterized clinically by fibrinous polyserositis, polyarthritis and septicaemia.

Etiology

G. parasuis earlier kinown as *Haemophilus parasuis* is a gram negative, polymorphic bacterium of Pasteurellaceae which need V- factor (NAD)

supplementation for growth. It can be grown on enriched chocolate agar and blood agar. *G. parasuis* is fastidious and its isolation from diseased animal is difficult due to antibiotic treatment. Fifteen serovar of *G. parasuis* is isolated based on genotyping.

Pathogenesis

The vasculitis plays an important role in the development of lesions in glasser disease but pathogenesis is not known. The clinical signs of disease can be due to primary pathogen or other pathogens like porcine reproductive and respiratory syndrome virus (PRRSV) or swine influenza virus (SIV) associated with primary disease. Presumably, there is a period of bacteremia since *G. parasuis* often localizes at multiple sites.

Clinical Signs

It mainly affects the pigs between age groups for 4-8 weeks but other age groups also affected.

In per acute cases, sudden death without clinical signs reported with petechiae in some organs.

In acute cases, typical clinical signs of high fever (106°F), abdominal breathing, severe cough, swollen joints, and CNS signs of trembling, paddling and lateral decubitus may be present.

In chronic cases, affected pigs have retarded growth due to severe fibrosis in thoracic and peritoneal cavity. Stress and immunosuppressive diseases like porcine reproductive and respiratory syndrome and porcine circovirus 2 infection may modify the prevalence of glasser's disease.

Post-Mortem Findings

In per acute cases only petechial lesions and fluid in thoracic and abdominal cavity without fibrins are present in organs while microscopic lesions like DIC and haemorrhages may be present. Acute systemic infection shows gross lesions such as fibrinous polyserositis, polyarthritis and meningitis. Besides, fibrinous exudates in pleura, pericardium, peritoneum, synovia, and meninges are reported. The cranioventral consolidation (suppurative bronchopneumonia) may be present in fibrinous pleuritis. Chronic cases are commonly having severe fibrosis in pericardium and pleura.

Diagnosis

The presumptive diagnosis can be made based on history of prevalence of disease, signs, and typical lesions of polyserositis and polyarthritis. The confirmatory diagnosis may be done by culture of *G. parasuis* from lesions affecting serous membranes, cerebrospinal fluid etc. PCR are for identifying *G. parasuis* from clinical specimens is also used for confirmatory diagnosis.

Prevention and Control

Prevention of gassers disease includes management modifications, control of concurrent disease like PRRSV, prophylactic antimicrobials and vaccinations. The diseased animals should be treated promptly with appropriate antimicrobial. Minimising stress and vaccination of breeding stock is useful in prevention of disease. Formalin killed bacterins are effective but does not give protection against all serovars.

7.3.3. Brucellosis

Swine brucellosis is a zoonotic disease caused by *Brucellas suis* and characterized by fever, retention of placenta and abortion (Bang's disease).

Etiology

Brucellosis in swine is caused by pleomorphic coccobacillus, non-sporing, non-motile intracellular organisms caused by *Brucella suis.* It may appear as singly, in pair, groups or short chains. *B. suis* has 5 biovars but infection in pigs mainly caused by biovars 1, 2 or 3.

Epidemiology

The brucellosis is prevalent throughout the world. Brucellosis in pigs is wide spread in Latin America, South east Asia and the Pacific region. It affects almost all species of animals and human being. The people who are in close contact of infected animals are especially prone for the infections like farmers, veterinarian, slaughter house workers etc. Brucella suis causes bacteremia which persist for up to 90 days and continue to shed for up to 3 years. Transmission of brucellosis to susceptible species and human being takes place through any secretion or excretion from the infected animals like aborted materials, milks, and milk products etc. Prevalence of *B. suis* in Nagaland and Punjab was 3% and 5.6%, respectively while in farm and stray pigs was 1.66% and 2.59%, respectively.

Pathogenesis

Brucellae organism is intracellular and has strong affinity towards lympho-reticular and reproductive systems which limits exposure to innate and adaptive immune responses. It prevents the organism from the effects of antibiotics, and exhibits clinical manifestations and pathology. Brucella evade intracellular destruction by restricting fusion of type IV secretion system-dependent brucella-containing vacuoles with lysosomal compartments. Besides, inhibition of apoptosis of infected mononuclear cells, and prevention of dendritic cell maturation, antigen presentation, and activation of naive T cells.

Clinical Signs

The clinical manifestation in pigs includes abortion, temporary or permanent sterility, orchitis in boar (usually unilateral), laminitis, hind limb paralysis, spondylitis and metritis. Abortions may occur early in gestation and may range from 0-80%.

Diagnosis

Serological test such as brucellosis card (rose bengal) test, serum agglutination test and complement fixations test are commonly used. Card test is more sensitive than other serological test. These tests have limitation in detection of brucellosis in individual animals hence entire herds or unit of herds should be vaccinated.

Treatment

Treatment is not recommended in pig infected with brucellosis.

Prevention and Control

The infected pig should be isolated and tested for brucellosis. Control of brucellosis in pig is mainly based on removal of infected breeding animals. Replacement pig introduced in herd should be purchase from farm free of babesiosis or they should be tested and isolated for 3 months before introduction in herd. There is no vaccine for swine brucellosis.

7.3.4. Pasteurellosis

Pasteurellosis in a highly contagious disease of pigs, caused by *Pasteurella multocida* and characterized by fever, coughing, dyspnoea, rhinitis and oedema.

Etiology

In pig pasteurellosis is caused by *Pasteurella multocida* which is a gram-negative bipolar organism and normal inhabitant of swine upper respiratory system. Atrophic rhinitis in pig is commonly associated with *P. multocida* types A and D whereas Pneumonia is mostly associated with type A. The disease is usually associated with complication of mycoplasma pneumonia. The swine influenza, Aujeszky disease, *Bordetella bronchiseptica* or *Haemophilus parahaemolyticus* may also cause changes in the lungs similar to *P. multocida.*

Epidemiology

The disease is widely prevalent in various species of animals all over the world. The disease has played a major role in huge economic losses in rural population due to high morbidity and mortality rate. Transmission is suspected through vertical and horizontal methods. Carrier swine is the major source of infection into most herds and transmission may take place by nose-to-nose contact.

Pathogenesis

The pigs in normal health are highly resistant to infection of Pasteurella sp. although they may act as carried of *P. multocida* in their naso-pharynx. When the bacteria colonises in deeper portion of lung during stress condition it causes chronic progressive pneumonia. The damage of micro-vasculature in lung by pathogenic strain causes fibrinous thrombi in alveolar capillaries and escape of fibrinogen into alveoli where it converted into fibrin. The necrotizing effect of *P. multocida* and other secondary bacterial infection causes formation of abscess in lung. Sometimes inflammatory reaction in pleura and pericardial sac is present. Fibrin on the pleura may form fibrinous adhesion which attach the lung to rib cage. *P. multocida* organisms are also present when infection of pneumonia causing agent *Actinobacillus sp.*, swine influenza, *Mycoplasma hyopneumoniae* present.

Clinical Signs

The clinical signs of pasteurellosis in swine vary in severity and are often associated with other types of pneumonia. The fever, coughing, dyspnea, and prostration are commonly reported in acute cases. Chronic cases of pasteurellosis are characterized by mild fever, a persistent cough, and a more marked dyspnea. In advance stage of pneumonia or extensive adhesions between lungs and rib cage often have a noticeable expiratory lift ("thump").

Post Mortem Findings

Pneumonic changes include the cranioventral lobes of the lungs and adjacent parts of the diaphragmatic lobes which is firm and dense. The pleural lesion areas may be granular and while regional lymph nodes are commonly enlarged. Microscopic lesions are those of a suppurative bronchopneumonia. Fibrin and Pasteurella organisms are often prominent in affected alveoli and, sometimes, in interalveolar capillaries.

Diagnosis

The pneumonic pasteurellosis in pigs may be diagnosed based on history, clinical signs, and the notable pulmonary lesions. The cases of pasteurellosis may be confirm by culture of lung lesions from a non-medicated pig. The diagnosis of pneumonic pasteurellosis also includes the consideration of predisposing factors especially environmental factors such as dusty or overcrowded conditions, excessive ammonia gas or poor ventilation.

Serology and mere isolation of *P. multocida* from nasal swab does not provide exact picture of disease as it may be *P. multocida* may be associated as secondary bacterial infection in severe other diseases causing pneumonia. The polymerase chain reaction (PCR) is useful technique to diagnosis pneumonic pasteurellosis and atrophic rhinitis.

Treatment

Early diagnosis, appropriate choice of antimicrobial is important in recovery from pasteurellosis. Many antibiotics like ceftriaxone, cefotaxime, tiamulin, lincomycin, penicillins, tilmicosin are effective in treatment. Antibiotics should be selected based on antibiotic sensitive test.

Prevention and Control

Prevention and control methods include all the measures which prevent the introduction of new strain of *P. multocida* into the herd. Farm practices like new animals before introduction in farm should be quarantined for 21 days and follow all in/all out system of raising swine to minimize the likelihood of pasteurellosis. Vaccine bacterins against *P. multocida* are available (Raksha HS) may be given at 2 months of age followed by at 6 months of age then annually.

7.4. DEFICIENCY DISEASES IN PIGS

7.4.1. Iron Deficiency Anaemia/Piglet Anaemia

Iron deficiency in pig is either primary or secondary deficiency. It is common in early life of sucking piglets under intensive management and in calves reared exclusively on milk feeding.

Etiology

Primarily deficiency is caused by iron deficiency in neonates reared on dam's milk as milk of most of the animals is deficient in iron content. Inadequate iron storage in the liver of piglet at birth causing latent iron deficiency in new born piglet. It results due to decrease production of uteroferrin after 70th day of gestation in sows when foetal growth is maximum. Copper and cobalt deficiency also causes anaemia as it acts as catalyst in haemoglobin formation. Iron (Fe) deficiency occurs in piglets confined in pens and concrete or other impervious floor and where sole diet is sow milk.

In farm animals, blood loss by any means also depletes "Fe" resources especially in lice infestation (ecto-parasitism), blood sucking worms (Haemonchus species, Bunostomum spp.). Addition of calcium carbonate in diet makes an insoluble complex with available iron where in "Fe" is not readily absorbed creating a secondary condition of Fe deficiency. High manganese content in soil also makes "Fe" unavailable.

Epidemiology

Iron deficiency is uncommon in adult animals because common feed and fodders contains enough iron. Iron deficiency anaemia is common in early life of farm bred piglets. Goats and cattle reared exclusively on grazing in some parts of India. Most common cause of anaemia are the secondary factors producing iron deficiency either through excessive drainage from the body or by interfering in the bioavailability of the iron for haemoglobin formation. Highest incidence occurs at about 3 weeks of age although piglet up to 10 weeks of age may be affected.

Pathogenesis

More than half of iron present in animal body is found as a constituent of haemoglobin and relatively small amount in myoglobin and certain enzymes which play a role in O_2 utilization.

Piglets are highly susceptible for "Fe" deficiency as piglets are borne with relatively low Hb concentration and there is physiological fall in Hb level in piglets to 4-5 gm of Hb/dl (Normal- 9-11 gm/dl) at about 8-10 days after their birth. Piglet are dependent on sow milk only for source of iron at this stage and milk being poor source of iron resulting in anaemia. Hepatic level of iron in piglets are considerably low. Piglets have relatively faster growth rate for e.g. they reach 4 to 5 times of their birth weight at the end of 3 weeks and 8 times of their birth weight at end of 8 weeks. The daily requirement of "Fe" during first few weeks of life is about 15mg/piglet while average concentration of "Fe" available through sow milk to individual piglet is about 1 mg. Hence, it leads to severe anaemia. Iron deficiency also leads to susceptibility of piglets to recurrent episodes of diarrhoea due to atrophy of parietal cells, decreased acid and chloride secretion which normally help in curdling of milk and assimilation. So, it leads to diarrhoea. Villous atrophy of small intestine leads to decreased absorption and consequently diarrhoea – responsible for less weight gain in growing piglets.

Clinical Findings

Diarrhoea is most encountered clinical feature. Normal colour faeces without any offensive/unpleasant smell present. Anaemic signs like severe dyspnoea, lethargy, tachycardia, pale mucosae, oedema of head and fore quarters giving puffed up appearance (Swollen face). Increased amplitude of heart beat even in mild exercise (thumping heart beat) reported.

Diagnosis

History of susceptibility of young age (d+2-3 weeks of age) and clinical symptoms of anaemia, pale mucous membrane, respiratory distress, tachycardia, thumping heart beats, recurrent episodes of diarrhoea.

Laboratory investigations showed low Hb (2-4 gm %) and TEC (3-4 million/Cu mm) count. The microscopic examination of RBC showed very characteristic microcytic, hypochromic type of anaemia. Serum iron level in cattle and sheep is 100-200 µg/dl of serum and in pigs 50-100 µg/dl of serum and less than these values indicate deficiency of iron.

Necropsy Findings

Pallor, thin and watery blood, and moderate anasarca along with cardiac dilatation and hypertrophy

Liver is enlarged and show mottled appearance and greyish yellow colour due to fatty infiltration

The mucosa of gastric fundus is shallower and less cellular and shows a pronounced decrease in parietal cells indicates decrease capacity of stomach to secret acid.

Treatment

- **Supplementation of Iron.**
  - Oral dosing: Ferrous sulphate @ 300 mg/day for 2 weeks is an effective therapy but it is less palatable and induces diarrhea due to irritation to gastrointestinal mucosa.
  - Parental route
    - ✓ Iron dextran complex (Imferon, 50 mg of elementary iron/ml) @ 200 mg (2-4 ml)/week for 3 weeks

- ✓ Iron sorbitol citric acid complex (Jactofer) is available as 50 mg/ml @ 100 mg/week for 3-4 weeks by i/m route.

- **Supportive Therapy**
  - ✓ Parenteral fluids containing 10% dextrose.
  - ✓ Good environment like moderate temperature
  - ✓ Good bedding

Prevention and Control

Dietary supplementation of iron: Sows can be fed with 2 gms of Fe/kg diet during late stage of pregnancy and upto 3 weeks post farrowing. Because it is observed that piglet ingests 20gm of sows faces per day which provide sufficient iron. Individual oral dosing of piglets 1.8% solution of ferrous sulphate is administered @ 4 ml/piglet as a dietary protocol for first 3-4 weeks. Alternatively, the sow udder can be painted with a solution containing 450 gms of $FeSO_4$, 75 gms $CuSO_4$, 450 gm of Sugar in 2 liters of water. Individual piglet gets required iron during suckling. Intramuscular injection of iron dextran/iron sorbitol complex 200 mg/Piglet on 3rd day of life and to be repeated after14 days.

7.5. PROTOZOAN DISEASE IN PIGS

7.5.1. Trypanosomiasis

Blood protozoan diseases commonly reported in pigs are trypanosomiasis and babesiosis. Trypanosoma infect all most all species of domestic and wild animals. Trypanosoma infecting pigs are *T. congolense, T. vivax, T. brucei, T. simiae, T. godfreyi* and *T. suis.* The prevalence of *T. simiae* is higher than other trypanosome species in pigs and it is highly pathogenic in domestic pigs. *T. brucei* identified in pig are zoonotic hence it is of public health concern. All species infecting pigs are salivarian type hence infective stage are released in the saliva. Transmission occurs when tsetse flies (Glossina sp.) and other biting insects (tabanids) ingest trypomastigotes in blood from infected pig. They convert into procyclic trypomastigotes and reproduce by binary fission in the midgut or proboscis of flies and after leaving it transform into epimastigotes. Epimastigotes in salivary glands will continue to multiply and transform into metacyclic form which are infective to the pig. In pig clinical signs include depression, anemia, fever, reduced fertility, enlargement of lymph nodes and acute death while signs of rapid emaciation, recurrent coughing, and diarrhea are present in piglets. Trypanosomes within the host will eat host's blood glucose, result in hypoglycemia hence infected pigs will be weak, lethargic, and emaciated. A presumptive diagnosis is made based on history and clinical findings of anemic, fever, and anorexia. Thin blood smear using Giemsa stain is used for detection of trypanosomes. Serology tests like ELISA and CATT are more suitable for herd screening and diagnosis. Molecular techniques like PCR are used for confirmatory diagnosis of trypanosome. Trypanosomiasis can be treated using medicine, Diminazen aceturate/Suramin or Trypamidium. Farm sanitation, application of insecticide and fogging to minimize vector (biting flies) and routine screening of herd to control the spread of disease which has major influences on animal health.

7.5.2. Babesiosis

Babesiosis is a tick-borne haemoprotozoan disease caused by *Babesia* spp., which affects wide range of vertebrates including human. The various species of animals affected by different species of babesia are Cattle with *B. bigemina*, *B. bovis*, dogs with *B. canis*, *B. gibsoni*, *B. rossi*, B. *vogeli* , cat with *B. felis*, horse with B. equi and B. caballi and small ruminants with B. motasi, *B. ovisare* the frequently affected species. The babesiosis infection in pigs is not very common and caused by *B. suis* and *B. tautmanni.* Swine babesiosis is rare but two outbreaks are reported in domestic pigs in Limpopo Province, South Africa and Zimbabwe. Babesiosis is worldwide in prevalent but is frequently reported in tropical and subtropical regions as climatic conditions are more conducive for ticks. The babesia organism is intra-erythrocytic which infect and multiply within RBC which ultimately result in lysis of erythrocyte. The damage of erythrocyte due to direct and indirect method result in hemolytic anemia, icterus, nephrosis and hemoglobinuria. The clinical signs like anorexia, anemia, lethargy, lameness, reluctance to stand, high fever, and abortion are frequently reported in pig. Diminazen aceturate is commonly used for the treatment of babesiosis in pigs.

7.5.3. Toxoplasmosis

Toxoplasmosis is a zoonotic disease caused by a protozoan parasite *Toxoplasma gondii* which is wide spread in swine worldwide. The ingestion of uncooked pork infected with *T. gondi* can be a source of infection for humans. The major source of infection of *T. gondii* are ingestion of contaminated food and water from cat faeces contaminated with Toxoplasma oocysts, ingestion of meat contaminated with tissue cysts and congenitally from an infected mother to her fetus. Wild and domestic cat act as definitive host for the parasites and excretes oocysts in their faeces. The toxoplasma commonly present in edible tissues of pigs and generally does not make pig ill. Nursing pigs generally show clinical signs which includes dead born piglets, sick, or turn out to be sick within 3 weeks after birth. Besides, fever, general weakness, diarrhoea and nervous signs while some remain clinically normal. Laboured respiration is the most common clinical sign of toxoplasmosis. Proper cooking, freezing (-12°F), or irradiation (100 krad) may kill toxoplasma in meat easily. Toxoplasma gondii may become innocuous by heating to an internal temperature of 67°C. The chances of infection in pigs may be minimize by Keeping cats out of the pig barns, feed, and water, removal of dead carcass immediately, use of rodenticide and avoid feeding of uncooked garbage to pigs.

7.5.4. Coccidiosis

Coccidiosis in swine is worldwide in distribution and commonly Eimeria spp. infects older pig and Isospora suis infects piglets. Eimeria species has low pathogenicity and do not infect the adult pigs. Adult pigs are infected in the small intestine most commonly *E. debliecki* and less frequently by *E. neodebliecki, E. perminuta, E. polita, E. porci, E. scabra, E. spinosa, E. suis*, and *E. szechuanensis. Eimeria debliecki* mainly infects the small intestine and *Eimeria neodebliecki* affects the mid portion of the jejunum. *Isospora suis* infection is prevalent in both intensive and extensive production in piglets. The sporulated oocysts are main source of infection in piglets when exposed to contaminated farrowing and rearing housing. If the infection is contracted in early stage of life, then it causes more severe disease. Piglets are mostly infected between 5-15 days of age and develop diarrhoea varying from creamy pasty consistency or white faeces to profuse watery yellowish diarrhoea with thriftiness. In older pig, clinical signs are mild and generally frothy, mucoid, watery diarrhoea with wasting present. The faecal examination of infected swine showed oocysts with sodium chloride or glucose solution. The pens and area must be cleaned thoroughly with effective disinfectant. The clinically infected piglets and pig in the same group should be treated with anticoccidial. Sulfonamides like Sulfamethazine (sulfadimidine) and sulfaquinoxaline in drinking water can be effective. Trimethoprim sulfonamide combination is also effective through oral or parenteral route. Amprolium @ 10–25 mg/kg, PO for 4–5 days in individual pig and @10 kg 25% premixed per ton of feed in group is effective in reducing the oocytes count in swine. Toltrazuril @ 20mg/ kg b wt once and monensin @ 100g/ton of feed is also effective in prevention of coccidiosis in swine.

7.5.5. Balantidiosis

Balantidiosis caused by Balentidia spp. is worldwide in distribution among various species of animals and human. The Balantidium reported from the pig are *B. coli* and *B. suis*. It is prevalent in tropical and sub-tropical countries as the climatic condition is suitable for their development and reproduction. Balantidium is generally harmless for pig but occasionally reported fatal in swine. Some pigs show moderate to severe diarrhoea while some may not exhibit any signs. All the age group of pigs are affected by balantidiosis. The infections generally occur through faeco-oral route as cysts and trophozoites pass in the faeces. But, only environmentally-resistant cyst can survive in environment and act as source of contamination to food and water. After ingestion, cysts hatch in the host intestine and releases motile form which begins to feed on starch grains, organic matter, and cell fragments. Terramycin @ 15 mg/kg body twice daily in concentrate feed is effective in clinical cases in pigs. Besides, Metronidazole and Diidohydroxyquine are effective in treatment of Balentidiosis.

7.6. ENDOPARASITES (INTERNAL PARASITES) IN PIGS

The internal parasites are mostly found in the digestive tract and use nutrients from the host to replicate and survive. Besides, few parasites are also reported in liver, kidney, lungs, and blood streams. In pig, the important parasites are the large white ascarids (*Ascaris suum*), whip worms (*Trichuris suis*) and stomach worms (*Hyostrongylus rubidus*) while in piglets thread worm (*Strongyloides ransomi*). Piglets become infected from the infected sow through transplacental route or through milk. The endoparasites in pigs and piglets are divided in three groups nematodes (round worm), thorny headed worms and tapeworms.

Most of the infection in healthy piglets or pig endoparasitic infection occurs through faeco-oral route. The common mode of transmission in pigs and piglets are keeping the pigs and piglets in unhygienic herd, lack of proper disinfection of drinking bowls and feeders, contact with other infected animals and grasses in premises. The parasites eggs enter the body of a cloven-hoofed animal known as cyst which in favourable condition to develop into worm. In piglets, it causes weight loss, anaemia, coughing, wheezes and pot belly. Lung worms (Metastrongylus spp.) in pig is caused by nematodes and earthworm act as intermediate host. Pig gets infection through ingestion of third stage lung worm larvae and mostly reported in domestic pigs reared outdoor. Five genera of stomach worm in pigs are *Trichostrongylid hyostrongylus* (common) and other four spirurids (Ascarops, Gnathostoma, Physocephalus, Simondsia) are less common. Light worm infestation is usually asymptomatic in pig while heavy worm infestation or host in poor health condition, these worms can cause weight loss, inappetence and anaemia resultant in "thin sow syndrome". *T. spiralis* is an encapsulated nematode affecting humans and domestic animals with high infectivity for pigs. Cyst of trichinella is commonly present in the striated muscle of pig and act as a source of infection to human being when consumed uncooked pork. Tapeworm (*Taenia solium*) is a flat and ribbon-like body having human or dog as primary host and pig as secondary or intermediate host. Pigs get infection through human feces that contain the parasite eggs and contaminate their fodder. Ingested eggs in intestine of pig develop into larvae followed by oncospheres, and eventually into infective tapeworm cysts commonly known as cysticercus

(*Cysticercus cellulosae*). Cysticercus is present in the muscle of pig hence it is known as "measly pork". Humans get infection through consumption of uncooked or under-cooked pork and develop into adult worm in the intestine. The treatment of nematode infection includes ivermectin@ 0.3 mg/kg b wt, Piperazine @ 200-400 mg/kg b wt orally, Albendazole @ 7.5-10 mg/kg b wt orally etc while for tape worm infection treatment includes praziquantel @ 5 mg/kg b wt orally. The pig should be dewormed at least at 4 months interval to minimize the internal parasitic load.

The worm infestation in pigs may be minimize by regular dispose of all fecal matter in the pig's habitat. The pet pigs over 20 pounds should be dewormed at eight weeks of age, then every two months. The daily cleaning of pens and treatment with disinfectants 2 times a week. The bedding in the pens is changed every day. The pen feeders and drinkers should be treated with boiling water once in 10 days. Besides, pens should be regularly whitewashed with lime and treated with solution of lye. The bedding should be cleanout from the pig pens. All raw vegetables and fruits before eating must wash, peel, or cook properly, especially those that has grown in fertilizer with pig manure. Public education regarding possible dangers of consuming raw and undercooked meat or pork, may reduce infection rates. Meat testing for the presence of cyst in pork of individual carcasses and monitoring of the herds. Larvae may be killed by the heating or irradiation of raw meat. Good nutrition may mount an immune response and maintain performance even in the presence of worm.

7.7. ECTOPARASITES IN PIGS

The ectoparasite are external parasites present over the body of pig and feed and reproduce on the body of pig. In pig, common ectoparasites are mites, lice, and ticks. These parasites not only damage the hide of pig but also transmit many life-threatening diseases. The annoyance associated caused by parasite infestation can cause slow weight gains and pig required more feed per kilogram of body weight gain compare to uninfected pig. The various clinical signs shown in pigs are scratching, bruises and hyperkeratosis. Ectoparasites in pig cause significant economic effects due to slow growth rate, decreased feed efficiency, and low of carcass value at slaughter. Mites present in pigs are *Sarcoptes scabiei var suis* infection commonly reported in pig and in acute infection causes head shaking, severe rubbing of body and allergic reaction leads to formation of red pimple all over the body. Chronic infection shows thick incrustation occurs in the ear, along the side of neck, elbow, and front parts of hock. These lesions consisting of high load of mites. *Demodectic phylloides* (Demodectic mange) is present in the hair follicle and it is relatively less important in pig as lesions are not much severe. The mites present outside the body are sensitive to exposure of sunlight and die in 24-48 hours. Contaminated bedding should be removed and premises should be cleaned with disinfectant. Quarantine of new comer and isolation of infected pig are helpful in preventing the infection in herd. The depopulation and repopulation of mange infected pig by segregated rearing and treatment with acaricides. The ivermectin @ 0.3 mg/kg b.wt s/c at weekly intervals or doramectin @ 0.3 mg.kg b wt in neck region until last skin scrapping examination is negative for the presence of mite.

7.7.1. Lice

Haematopinus suis has piercing and sucking mouthparts and males are slightly smaller than female. The lice are greyish-brown colour with brown to black markings and become blueish in appearance after feeding. Transmission occurs through direct mode. Lice cannot survive for two to three days away from lice body as it is host specific. It is usually present all over the body but especially in the wrinkles of skin around the neck, flanks and in the inguinal region. Lice usually causes anaemia, slow growth rate and feed efficiency in young pigs. Treatments of lice infestation include application of different acaricides like flumethrin, fipronil, ivermectin in various formulation viz., sprays, pour-on and injections. These acaricides can cure lice if administered two times at 10-14 days intervals but not effective against eggs of lice.

Tick

Ticks are acaricides present on the body of pig and are solely dependent on pig for their food as blood. The ticks are present all over the body of pig but preferred site is soft part of body like the common tick present on the body of pig is *Rhipicephalus sanguineus.* The tick act as vector of various disease in pigs including ehrlichiosis, lyme disease, anaplasmosis etc. These diseases have very detrimental effect on the body of pig and it may cause death of pig. The various acaricides like flumethrin, fipronil, ivermectin in different formulations viz., sprays, pour-on and injections are effective to knock down the ticks. The premises of pig shed should also be disinfected with acaricides at high concentration like deltamethrin @ 4-5 ml/1 liter of water.

CHAPTER 8

VISUALIZING DISEASES: POSTMORTEM INSIGHTS INTO PIG PATHOLOGY

Dr. Kanika Kalai

8.1.INTRODUCTION

Pigs are important contributors to the global livestock industry and serve as a primary source of meat and related products in the world. In India, pigs are considered as a food for backward people but with advancing technology and with ever increasing population, the demand for pig meat or pork has also rises. In North East India, pork is the delicacy of almost all non-vegetarian dishes. That is the reason behind the pig industry gaining popularity in this part and attracting many young entrepreneurs to invest their time and money. However, susceptibility of pigs to various diseases poses significant challenges to profitable pig farming and also posing potential food security hazards. Understanding pig lesions associated with diseases is crucial for diagnosis of disease and its control and prevention. This chapter will provide an overview of the lesions occur in various disease conditions in pigs.

Lesions can be defined as any pathological changes, alterations or abnormalities that occur in the organs, tissues or systems of pigs as a result of disease or infection. These lesions can be manifested in various forms from ultrastructural changes to grossly visible abnormalities that can be observed during post-mortem examinations or microscopically detectable alterations in tissue samples during histopathological examination.

DOI: 10.1201/9781003749844-8

Diseases affecting pigs can be originated from multiple sources, including bacteria, viruses, parasites, fungi, metabolic abnormalities and nutritional imbalances. Some of these diseases are highly contagious and spread very fast within swine populations and some may be endemic or sporadic. These lesions provide valuable diagnostic clues aiding the veterinarians and researchers in pinpointing the specific disease responsible for the observed abnormalities.

The consequences of failing to recognize and identify the lesions associated with diseases can cause uncontrolled outbreaks of infectious diseases which may lead to high mortality rates, reduced growth rates, decreased reproductive performance and compromised meat quality.

8.2. POST MORTEM EXAMINATION IN PIGS

A post-mortem examination, also known as a necropsy in the case of animals, is a systematic examination of dead animals by opening the carcass to determine the cause of death and evaluate its overall health before death.

A post-mortem examination of a pig is conducted by a traned veterinarian or a pathologist. It can provide valuable understanding into the health and potential diseases affecting swine populations.

External Examination

The examination begins with a thorough external assessment. The examiner checks the pig's body for any visible abnormalities, injuries or signs of disease. This includes observing the pig's skin, external features, the status of rigor mortis, any abnormal discharge from the natural orifices etc. in the body.

Internal Examination

Opening the Body: An incision is made along the midline of the pig's abdomen to open the internal organs. The skin and underlying tissues are carefully cut and peeled back. The mucous and serous membranes, muscles are observed carefully.

Observation of Organs: After opening of the abdominal cavity the internal organs are exposed systematically. The examiner observes the colour, texture and overall condition of each organ. Any abnormalities or lesions are to be noted.

Chest and Thoracic Cavity: The chest cavity is opened to examine the heart and lungs. This is important to check for respiratory diseases or cardiovascular issues. Hydropericardium, pneumonia, pulmonary oedema etc are important lesions to be noticed. Cardiac failure may occur without changes in other organs in porcine stress syndrome, a genetic disorder.

Pneumonia may be observed in *Mycoplasma hyopneumoniae* infection with chronic lungs lesion and in Glasser's disease (*Haemophilus parasuis*). Pleuritis with lesions in lungs are observed in *Actinobacillus suis* infection.

Abdominal Cavity: Stomach, small and large intestine are checked properly for any abnormalities. To avoid soiling of the visceral organs by intestinal contents, the digestive tract should be opened at the last. Gastric ulcers can be found in stress, infection and other conditions. In swine vesicular disease caused by Picorna virus, the lesions are found in mouth cavity along with feet. Thickening of the small intestine and ileitis observed in porcine intestinal adenomatosis caused by the bacterium *Lawsonia intracellularis*.

Liver is checked for congestion, necrosis and other lesions. Liver abscess can be formed by *Actinobacillus pleuropneumoniae.* Hydatid cyst, *Ascaris suum, Cysticercous cellulosae, Cysticercous tenuicollis* can be found in liver. Tiny focal cyst may be present without posing much hampering to the liver function but numerous nos. of cysts with diffuse distribution can compromise the structure and function of liver.

Pelvic Cavity: Kidneys, urinary bladder should be checked. Nephritis, hydronephrosis, dehydration, lithiasis, haemorrhage are important lesions in kidneys. Haemorrhages can be observed after peeling the capsules. Also, capsules will be not be easily peeled off or come with tissues in case of chronic nephritis. Polycystic kidneys can be found which is a hereditary condition characterized by the development of multiple fluid-filled cysts in the kidneys without development of lesions in other organs.

Brain and Head Examination: In some cases, especially when neurological symptoms are present (Rabies), the skull may be opened to examine the brain for abnormalities or diseases and for collection of brain tissue for further study.

Recording of Findings: Detailed notes and photographs are to be taken to document the findings throughout the examination as this documentation is crucial for diagnostic purposes and especially important in veterolegal cases.

Sampling: Tissue samples should be collected from organs where gross lesions are observed along with parts of normal tissue and also from the tissues where no observable changes are detected particularly if there are suspected disease conditions. These samples are collected either aseptically for microbiological analysis or in 10% formalin for histopathological analysis. Materials for isolation of virus should be collected in 50% glycerine phosphate buffer.

Final Report: Once the examination is complete, a final report is prepared, summarizing the findings and conclusions. If a cause of death or disease is identified, it is documented.

8.3. IMPORTANT DISEASES WITH RELATED LESIONS IN PIGS

8.3.1. Classical Swine Fever

Classical swine fever (CSF) or "hog cholera" is a highly contagious viral disease affecting domestic and wild pigs. The disease is caused by the Classical Swine Fever Virus (CSFV), a member of the genus Pestivirus under the Flaviviridae family (WOAH). CSF can result in various clinical signs and lesions in affected pigs. Lesions associated with classical swine fever are mainly because of the direct effect on the vascular system especially due to capillary degeneration.

Gross Lesions: Haemorrhages in internal organs are commonly observed in the kidneys, spleen and lymph nodes. Swollen and enlarged lymph nodes are a common finding in classical swine fever. The lymph nodes may be congested and haemorrhagic. Lesions in lymph nodes are evident in more than 80% cases of swine fever, where the pig died due to this disease. Infected pigs may exhibit necrotic (localised death tissue) changes in the tonsils along with haemorrhage. The spleen may show infarcts (areas of tissue death which is usually a coagulative type of necrosis due to occlusion of the supplying end artery) which appear as raised nodules in the borders of the spleen. Haemorrhages may also be present. Infarction in the spleen is considered as one of the pathognomonic lesions for swine fever. Haemorrhagic and necrotic changes may occur in the gastrointestinal tract particularly in the Peyer's patches of the small intestine. Button shaped ulcers may also form in the intestine, usually in the colon in chronic infection. Pneumonia-like changes may be observed in lungs. In the kidneys sharply demarcated petechial to ecchymotic hemorrhagic spots develop which are a very prominent lesion after removal of the capsules. The changes give rise to the condition which is described as "turkey egg appearance" of the kidneys due to its resemblance with the eggs of turkey birds. In severe cases CSFV can affect the central nervous system leading to neurological signs. Lesions like congestion may be present in the brain and affected pigs may show symptoms such as incoordination, tremors and paralysis. In pregnant sows there may be metritis, placentitis and other lesions. Erythema (redness), cyanoses, hemorrhages etc may be observed in skin. Severity and distribution of lesions can vary depending on the animal and the strain of the virus. For confirmatory diagnosis of the disease laboratory tests, immunological test and molecular detections are necessary.

If a pregnant sow gets exposed to the virus during gestational period, then foetal lesions like foetal mummification, anasarca, ascites, still birth, cerebellar hypoplasia and neonatal death are seen along with abortion of the foetus.

Histopathological Lesions

Lymphoid depletion may be seen by reduction in the number of lymphocytes in lymph nodes, spleen and other lymphoid organs. Lymphoid tissues may show areas of necrosis, particularly in the lymph nodes. Necrotic areas can be observed in the white pulp of the spleen. Necrotic lesions in the tonsils may observed. Petechial (pinpoint) or ecchymotic haemorrhages may be present in the kidneys. Necrotizing and haemorrhagic enteritis along with ulcers can be seen. In severe cases, there may be interstitial pneumonia with inflammatory infiltrates in the lungs. The virus can affect the central nervous system leading to encephalitis and congestion and other inflammatory changes can be observed under microscope. Inflammation of blood vessels like hyalinization, fibrinoid necrosis, DIC (Disseminated intravascular coagulation) leading to haemorrhages may occur particularly in smaller vessels of visceral organs.

8.3.2. African Swine Fever

African Swine Fever (ASF) also known as "warthog disease" and "African pig disease", is a highly contagious and often fatal viral disease affecting domestic and wild pigs caused by a double-stranded DNA virus belonging to the Asfarviridae family. The virus is named African Swine Fever Virus (ASFV). It is more acute and severe in nature than the classical swine fever disease. The virus targets macrophages, causing their destruction and affecting the immune system's ability to mount an effective response.

Gross Lesions: Gross lesions include haemorrhage and congestion in the skin, mucous membranes and internal organs. Enlarged, congested lymph nodes and spleen are seen. Edema and necrosis occur in various tissues. Haemorrhagic lymphadenitis is a characteristic finding in this disease. Serosanguinous transudate may accumulate in body cavities in this disease.

Histopathological Changes: Microscopic examination often reveals widespread haemorrhages in various tissues, including the skin, mucous membranes and internal organs. These haemorrhages contribute to the characteristic clinical sign of reddening or discoloration of the skin. The virus targets lymphoid tissues, leading to severe changes. Lymph nodes and spleen may exhibit lymphoid depletion causing significant reduction in the number of lymphocytes. Lymph nodes often show haemorrhagic lymphadenitis, characterized by haemorrhage within the lymph nodes. This is a key histopathological feature of ASF. The spleen undergoes significant changes including congestion, enlargement and haemorrhagic lesions. Necrotic changes develop in the spleen, lymph nodes and other affected organs. Necrosis is often associated with the destructive impact of the virus on host cells. Vasculitis and haemorrhage observed in different tissues and organs. Edema or fluid accumulation may be present in affected tissues. Edema is often seen in conjunction with hemorrhages and necrotic changes. In severe cases, the respiratory and gastrointestinal tracts may show lesions such as interstitial pneumonia and necrotizing enteritis, respectively. The respiratory lesions are more common in African swine fever than in classical swine fever. The skin may exhibit hemorrhages, necrosis and other changes. These cutaneous alterations contribute to the characteristic reddening or cyanosis observed in affected pigs. The bone marrow may show depletion of hematopoietic cells, impacting the production of blood cells.

8.3.3. Foot And Mouth Disease

Foot-and-Mouth Disease (FMD) is a highly contagious viral disease affecting cloven-hoofed animals, including pigs. It is caused by the Foot-and-Mouth Disease Virus (FMDV), a member of the Picornaviridae family. Infected pigs exhibit a sudden onset of high fever with formation of vesicles (fluid-filled blisters) in various parts of the body.

Gross Lesions: There may be presence of vesicles or ulcers on feet, snout and inside the mouth. When the vesicles get ruptured then painful ulcers are developed. The most characteristic lesion in pigs is erosions and ulcers at the coronary bands of the feet. Poor body condition due to anorexia caused by pain due to the presence of vesicles and ulcers and also due to the lameness caused by the lesions in coronary band.

Histopathological Lesions

Vesicles found in the epithelium of the skin and mucous membranes show acantholysis (loss of intercellular connections between epithelial cells) and ballooning degeneration of cells. Ruptured vesicles lead to the formation of ulcers and erosions in the affected tissues, particularly in areas such as the feet, mouth and snout. Ulcers are characterized by loss of epithelium and may be associated with inflammatory infiltrates. The ulcers at the coronary bands of the feet involve disruption of the epithelial layer and inflammatory responses. The tongue and oral mucosa reveal similar features of epithelial disruption, acantholysis and inflammation. Necrosis or cell death may be present in the affected epithelial cells and underlying tissue along with vasculitis. Histopathological changes can vary depending on factors such as the stage of the disease, the strain of the virus and the immune response of the host.

8.3.4. Porcine Reproductive And Respiratory Syndrome (PRRS)

It is an immune suppressive viral disease in pigs caused by the PRRS virus (PRRSV) which belongs to the family Arteriviridae. Respiratory and reproductive systems are mainly affected in this disease.

Respiratory Organs: Lungs are mainly affected and cause interstitial pneumonia characterized by inflammation between the alveoli. The interstitial pneumonia varies in distribution, ranging from multifocal to lobular to diffuse. Lungs may appear mottled and tan. The extent of these changes can vary significantly. Swollen, tan and oedematous or cystic lymph nodes are common. Microscopic lesions may include nonsuppurative interstitial pneumonia, mild non-suppurative encephalitis, myocarditis, rhinitis and potential depletion of germinal centres in lymph nodes (WOAH).

Reproductive Organs: In most cases no discernible lesions are noted in the reproductive tract. Some exhibit uterus lesions. The uterus may be enlarged and oedematous due to the reproductive failures associated with PRRSV. Abortion occurs in pregnant sow at any stage of gestation period but mostly observed in late trimester. There may be stillbirths, mummified foetuses and parturition of weak piglets. Arteritis and haemorrhage in the umbilical cord may be observed.

8.3.5. Porcine Circovirus (PCV)

It is a small, non-enveloped virus belonging to the Circoviridae family. There are two main types of porcine circovirus: PCV1 and PCV2. PCV1 is generally considered non-pathogenic and is commonly found in pig populations without causing significant disease. On the other hand, PCV2 is associated with several clinical syndromes collectively known as porcine circovirus-associated diseases (PCVAD). PCV2 is documented as a major pathogen in the swine industry worldwide and can lead to significant economic losses due to decreased productivity, increased mortality and treatment costs. It is now reported from various parts of India too. PCVAD encompasses a range of clinical manifestations, including post-weaning multisystemic wasting syndrome (PMWS), porcine dermatitis and nephropathy syndrome (PDNS), porcine respiratory disease complex (PRDC) and reproductive failure.

Post-Weaning Multisystemic Wasting Syndrome (PMWS): PMWS is characterized by weight loss, lethargy, respiratory distress and enlarged lymph nodes in pigs post-weaning. It is considered one of the most economically significant manifestations of PCV2 infection.

Porcine Respiratory Disease Complex (PRDC): PCV2 is often implicated as a component of PRDC. It is a multifactorial respiratory disease syndrome in pigs involving interactions between viral, bacterial and environmental factors.

Reproductive Failure: PCV2 infection has been associated with reproductive disorders in sows including abortions, stillbirths, mummified foetuses and infertility. Infected boars are the common source of infection to the sows and play very important role in transmission of the disease and outbreak in our common management and breeding system of pigs in villages.

8.3.6. Swine Influenza

Swine influenza is a highly contagious viral infection of pigs. All infected pigs may not show any clinical signs but can spread the disease very rapidly. It is caused by influenza A viruses, under *Orthomyxoviridae* family. The most common subtypes that infect pigs are H1N1, H1N2 and H3N2.

Gross Lesions: The lesions associated with swine influenza in pigs are typically found in the respiratory tract and can vary in severity. Dark red or purple discolouration with consolidated areas can be seen in lungs. These consolidated areas can sometimes be distributed in a multifocal or diffuse pattern, reflecting the nature of the infection. Bronchitis and bronchiolitis are also common. Oedema and hyperaemia may be observed in the affected lung tissues. The airways may contain a mucopurulent exudate which is a mixture of mucus and pus.

Histopathological Changes: Histopathological examination of lung tissues may reveal lesions such as necrotizing bronchiolitis and interstitial pneumonia. These changes indicate the destructive impact of the virus on the lung parenchyma.

8.3.7. Swine Pox

Swinepox is a sporadic viral disease of domestic and wild pigs that mainly occurs during the rainy season. The disease is caused by Swine pox virus under the Poxviridae family. The virus is resistant to environmental degradation and persists in scabs for up to a year. It is generally a mild disease and is not considered as a significant threat to the swine industry.

Gross Lesions: Swinepox is characterized by the development of skin lesions in affected pigs. These lesions typically appear as raised, firm and round papules or nodules on the skin. The lesions can range in size from a few mm to a cm or more in diameter. The lesions often start as red spots and progress to form raised, firm and round papules. They may have a central depression or dimple, giving them a characteristic appearance. Lesions can be scattered or clustered and can occur on various parts of the pig's body, including the abdomen, flank, ears and snout. However, they are most commonly observed on the ears. The lesions may develop a crust or scab gradually and can eventually heal.

Histopathological Lesions: Characteristic changes include epidermal hyperplasia, balooning degeneration of keratinocytes, intracytoplasmic inclusion bodies (Bollinger body) in kertinocytes and infiltration of inflammatory cells in the affected areas.

8.3.8. Transmissible Gastroenteritis (TGE) and Porcine Respiratory Coronavirus (PRCV) and Porcine Epidemic Diarrhoea (PED)

TGE is a contagious acute disease of pigs and affects all age groups. Mortality is more in young piglets under 2 weeks of age. The disease is caused by TGE virus under the coronaviridae family. Chronic form or the endemic form of the disease is known as porcine respiratory coronavirus (PRCV) infection. There is less mortality and severity of signs in this form but morbidity is high. Lesions are not very specific in this disease. In PED, the primary lesions are observed in the small intestine. Other organs such as the stomach and large intestine may also show signs of inflammation. In the small intestine villous atrophy occurs as the virus primarily targets the villous epithelial cells of the small intestine leading to severe atrophy of these structures. This results in a loss of absorptive surface area which leads to diarrhoea and malabsorption. Hyperaemia and enteritis observed in affected parts of the small intestine. Ultimately leading to death of affected animal.

8.3.9. Swine Erysipelas

Swine erysipelas is a bacterial disease caused by the bacteria *Erysipelothrix rhusiopathiae.* Lesions associated with swine erysipelas can manifest in different forms depending on the severity of the infection and the specific organ systems involved.

Gross Lesions: One of the classic signs of swine erysipelas is the appearance of diamond-shaped lesions on the skin. These lesions are raised, well-defined and have a characteristic diamond or rectangular shape. The lesions can be anywhere on the pig's body but are often observed on the back, neck and hindquarters. The bacteria have a predilection for the blood vessels and severe cases may result in damage to the blood vessels in the ears. This can lead to a condition known as "diamond skin disease,". The ear tips become necrotic (dead tissue) and may curl. It can affect the joints leading to arthritis. Affected joints may be swollen, hot to the touch and painful. In severe cases, the bacteria can enter the bloodstream and affect the heart valves leading to endocarditis. Vegetative endocarditis is the common feature. This can result in heart murmurs, lameness and sudden death. The disease can be manifested in acute and chronic form. The acute form of swine erysipelas may result in sudden death with minimal external lesions.

8.3.10. *Escherichia coli*

Escherichia coli (*E. coli*) is a type of bacteria that naturally resides in the intestines of animals and humans. Some pathogenic strains of *E coli* include: ETEC, Shiga like toxin producing *E coli,* Shiga toxin producing *E coli,* EPEC etc.

The lesions produced by various pathogenic strains of *E. coli* in pigs can vary based on the specific strain and the organs or tissues affected.

Enterotoxigenic *E. coli* (ETEC): ETEC primarily affects the small intestine. The bacteria attach to the intestinal lining and produce enterotoxins, leading to villous atrophy, epithelial cell damage and malabsorption. Lesions may include congestion and edema in the small intestine.

Edema Disease (ED): Edema disease is caused by specific strains of *E. coli* that produce a Shiga-like toxin. It primarily affects post-weaning piglets. Edema disease is characterized by lesions associated with vascular damage. Affected tissues particularly in the brain may show edema (fluid accumulation), congestion and hemorrhages. The small intestine may also exhibit congestion.

Enteropathogenic *E. coli* (EPEC): EPEC strains cause attaching and effacing lesions in the small intestine. This involves the bacteria attaching to the intestinal epithelium, effacing microvilli and causing localized damage. Lesions may include mucosal thickening and a characteristic "pedestal" formation at the site of bacterial attachment.

Shiga Toxin-Producing *E. coli* (STEC): STEC strains can produce toxins that lead to systemic effects. Hemorrhagic colitis characterized by mucosal hemorrhages and inflammation are seen in the intestine. Kidney damage may occur leading to lesions such as glomerular congestion and necrosis.

Neonatal *E. coli* Infections: In neonatal piglets, *E. coli* infections can lead to systemic disease. Lesions may be observed in multiple organs, including the lungs, liver, spleen and joints. In severe cases fibrinous polyserositis with fibrin deposits on the surfaces of various organs may occur.

The severity and distribution of lesions can vary and some strains may cause a combination of lesions. Additionally, lesions may not be limited to the gastrointestinal tract as systemic effects that can occur in certain infections.

8.3.11. Salmonellosis

Salmonellosis in pigs is caused by bacteria belonging to the genus Salmonella and it can result in various clinical manifestations including lesions in affected animals. Some pigs may carry the bacteria asymptomatically acting as carriers and shedding the bacteria in their faeces. In some cases, Salmonella can enter the bloodstream leading to septicaemia. This can result in disseminated lesions throughout various organs, including the liver and spleen. The severity and location of lesions can vary depending on the strain of Salmonella involved, the age and health status of the pig and other factors.

Gross Leisons: Primarily congestion, hemorrhage and inflammation observed in small intestine. In severe cases, the inflammation can extend to the cecum and colon, leading to typhlocolitis or commonly referred as enterocolitis. This may be associated with necrotic areas in the intestinal lining. Enlarged and congested mesenteric lymph nodes may be observed indicating the involvement of the lymphatic system in the spread of the infection. Some strains of Salmonella can cause joint infections leading to polyarthritis. Joints may be swollen with accumulation of fluid within the joint space along with other inflammatory changes. Severity of joint lesions may vary but piglets usually suffer more. Pneumonia may be observed in systemic salmonellosis.

8.3.12. Greasy Pig Disease

The disease is also known as exudative epidermitis, is a bacterial skin infection that primarily affects piglets. This condition is caused by strains of the bacteria *Staphylococcus hyicus*. Greasy Pig Disease can result in skin lesions, typically characterized by the development of a greasy and exudative (oozing) material on the skin. The severity of the disease can vary and it is more commonly seen in intensive pig farming systems. The initial stage of Greasy Pig Disease often involves areas of reddened skin. These areas may be warm to the touch and are indicative of inflammation. As the disease progresses the affected areas become moist and greasy. The skin appears shiny and may have an oily or greasy texture. Then exudation or discharge of yellowish fluid from skin may accumulate on the skin surface. Alopecia may occur in affected areas leaving the skin exposed. The hair loss contributes to the greasy appearance of the skin lesions. The condition is often associated with a foul odour emanating from the affected areas. This odour is a result of the bacterial infection and the breakdown of tissues. The lesions are commonly found on specific body parts, such as the back, sides and flanks of piglets. These areas are particularly susceptible to the development of lesions. In some cases, the exudate may become dry and form crusts or scabs on the surface of the lesions. This crusting can contribute to the overall appearance of the affected skin. In severe cases secondary bacterial infections may occur leading to more extensive lesions and complicating the disease. Tissue sample from the junction of affected and unaffected layers of epidermis and dermis should be collected in formalin for histopathologic examination. The pathognomonic histologic lesion include hyperkeratosis, parakeratosis (retention of nucleus in mature kerationacyte), dermatitis etc.

8.3.13. Brucellosis

It is a bacterial infection caused by *Brucella suis.* In breeding sows, brucellosis can lead to reproductive problems such as abortion, stillbirths and the birth of weak piglets. Lesions associated with reproductive involvement may include inflammation of the placenta (placentitis) and uterine infection. *Brucella suis* has an affinity for the joints and can cause arthritis and osteomyelitis (inflammation of the bone). In affected joints, there may be signs of inflammation, swelling and lameness. It can also lead to enlargement and inflammation of lymph nodes. Lymph nodes may be palpable and exhibit changes such as congestion and necrosis. In severe cases, brucella infection can spread to internal organs such as the liver and spleen. Hepatitis (inflammation of the liver) and splenomegaly (enlargement of the spleen) may be observed. In male pigs, brucellosis can cause orchitis (inflammation of the testicles) leading to infertility and testicular atrophy. Although less common, brucellosis can affect the central nervous system causing neurological signs. Lesions in the brain and spinal cord may be observed in affected animals. Brucellosis can cause inflammation of the mammary gland (mastitis) in sows. In advanced cases, brucellosis can cause generalized inflammation and lesions throughout the body.

8.3.14. *Streptococcus suis* Infection

S. suis plays a crucial role as a significant swine pathogen causing a diverse array of diseases including meningitis, arthritis, septicaemia, endocarditis, encephalitis, abortions, polyserositis and bronchopneumonia. *Streptococcus suis* is a major cause of meningitis in pigs characterized by inflammation of the membranes covering the brain and spinal cord. Lesions associated with meningitis may include congestion, haemorrhage and inflammation of the meninges. Joint involvement is common in *S. suis* infections leading to arthritis. Affected joints may show signs of inflammation, such as swelling and redness. There may be purulent (pus-filled) arthritis causing joint damage in severe cases. Widespread congestion, haemorrhage and damage to various organs occur in case of septicemia when the organism reach the bloodstream. Endocarditis especially vegetative endocarditis observed in many cases. Encephalitis or inflammation of the brain may occur in in this condition. Lesions in the brain may include areas of congestion, haemorrhage and inflammation. Metritis and placentitis leading to abortion may occur in some cases. Polyserositis with accumulation of fluid and inflammation in the pleura, pericardium and peritoneum may observed. Lesions in the lungs may include areas of consolidation, inflammation and pneumonia.

8.3.15. Methicillin Resistant *Staphylococcus aureus* (MRSA) Infection

Methicillin-resistant *Staphylococcus aureus* (MRSA) is a type of bacteria that has developed resistance to commonly used antibiotics including methicillin and other beta-lactam antibiotics. MRSA infections can occur in various animals including pigs and also in the human population. Swine can serve as carrier of MRSA means pigs can be colonized with MRSA without showing clinical signs of illness. The bacteria may be present in the nasal passages, skin and other mucous membranes. MRSA in pigs can be transmitted between animals and potentially to humans. Close contact with colonized pigs as well as contact with contaminated environments poses a risk.

Lesions: MRSA colonization in pigs usually not lead to clinical disease. Although few lesions may be observed in infected animals. There may be formation of abscesses, pustules or other inflammatory reactions in skin and other soft tissues. In some cases, the respiratory infections may lead to development of pneumonia. MRSA can cause joint infections in pigs leading to arthritis. Swollen and inflamed joints may be observed. Mastitis (inflammation of mammary glands) may also observed in sows.

Prevention and control: Preventive measures in swine farming include maintaining good hygiene practices, implementing biosecurity measures and practicing proper antibiotic use to prevent the development of antibiotic resistance.

8.3.16. Parasitic Infestation

Parasites play a very important role in pig health and some parasites also have important zoonotic implications impacting economic and public health. These parasites not only affect the well-being of pigs but also present potential health hazards for humans who come into contact with infected animals or their byproducts. Therefore, understanding and effectively managing parasites in pig populations are essential for maintaining both animal welfare and safeguarding public health and economic interests. Pigs can be infested by all types of parasites. Some important parasites of pigs are:

Ascaris suum: This roundworm infects the intestines of pigs and can lead to reduced growth rates, poor feed conversion and occasionally intestinal blockages.

Trichuris suis: Also known as the whipworm infects the large intestine of pigs and can cause diarrhoea, weight loss and anaemia.

Strongyloides: This nematode parasite affects the small intestine of pigs causing diarrhoea, weight loss and reduced growth rates.

Coccidia: Several species of coccidia specially Isospora can infect pigs, causing diarrhoea, dehydration and sometimes death particularly in young piglets.

Toxoplasma gondii: Pigs can become infected with this protozoan parasite through ingestion of contaminated feed or water. While it may not cause severe symptoms in pigs it can be a concern for human health if the meat is consumed undercooked.

Sarcoptes scabiei: This mite causes sarcoptic mange, a highly contagious skin disease characterized by intense itching, hair loss and skin lesions.

Gastrointestinal tapeworms: Tapeworms such as *Taenia solium* and *Taenia saginata* can infect pigs leading to reduced growth rates and occasionally causing cysticercosis in humans if the meat is consumed undercooked.

Lice: Various species of lice can infest pigs, causing irritation, skin damage and reduced productivity.

Ticks: Several species of ticks can parasitize pigs, leading to irritation, skin damage and potential transmission of diseases such as African swine fever and other tick-borne diseases.

Liver flukes (*Fasciola hepatica*): While more commonly associated with ruminants, pigs can also be infected by liver flukes leading to liver damage and reduced productivity.

The initial step to diagnose the parasitic infestations is a careful visual examination of the carcass and organs. Adult worms, cysts or lesions are evident in the intestines, liver, lungs and other organs where parasites commonly reside (Figures No. 1-5). Sometimes laboratory examination and molecular tests are necessary to confirm a parasitic infestation.

8.4. CONCLUSION

For gaining a comphrehensive perspective, it is said that one needs to see the forest as a whole, as well as an individual tree. Likewise, examination of an individual diseased pig and analysis of the whole herd is very important to diagnosis of a disease. Along with lesions observed during the post mortem examination and histopathological examination, history and clinical findings are also equally important in diagnosis of a disease condition. History includes age of pigs affected, duration of clinical signs, morbidity rate, mortality rate, treatments administered, response to treatments and any other important information regarding previous diagnoses or disease in the affected group of animals. This chapter has explored various lesions that serve as visual indicators for different diseases and help in quick diagnosis aiding for effective management and prevention of any major outbreaks to check any economic losses.

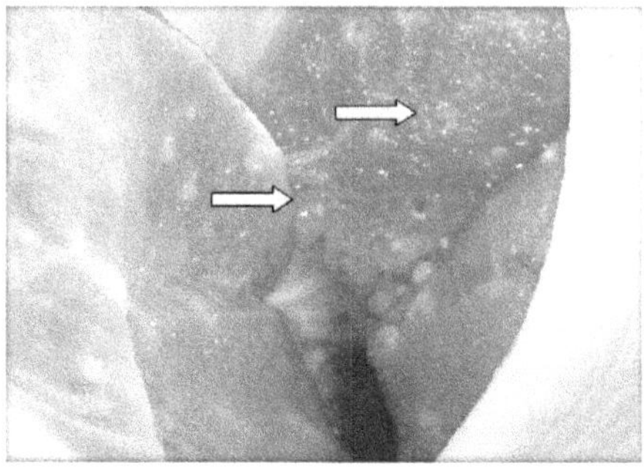

Measly pork (*C cellulosae*, intermediate stage of *T solium* in muscle)

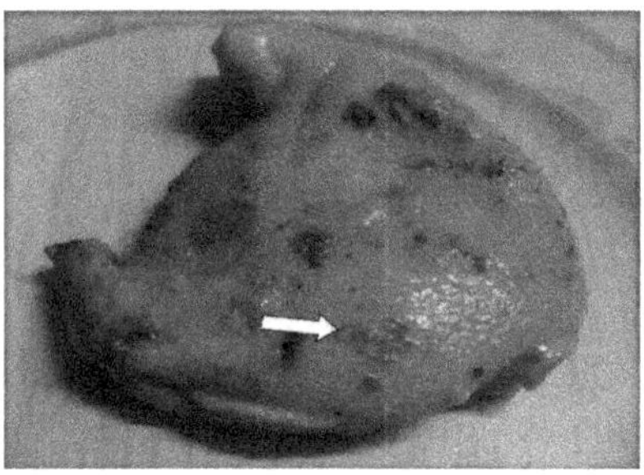

Small round worm in the stomach

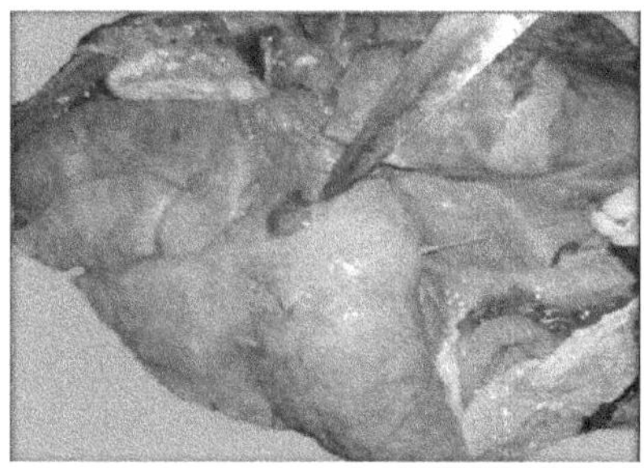

Hydatid cyst in lungs

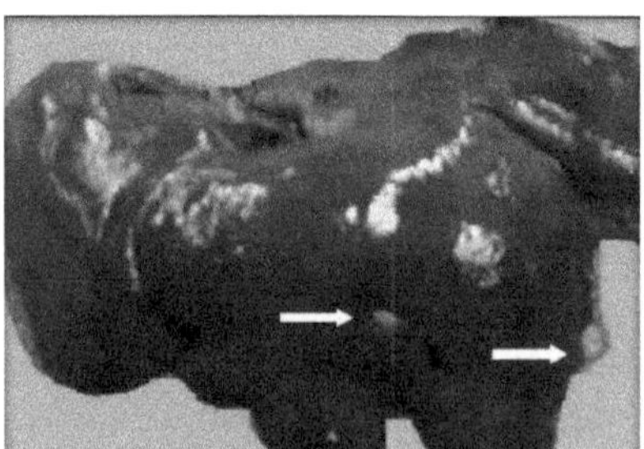

Hydatid cyst in liver

Milk spots (due to visceral larval migrants) and cyst in liver

The intricate study of pig lesions in various diseases unveils a fascinating journey into understanding the dynamic interactions between health and affliction in swine. This enhances our diagnostic capabilities and contributes to the broader field of veterinary science. By recognizing the nuances of lesions in different diseases, we pave the way for more effective preventive measures and targeted interventions, ultimately striving for improved swine health and well-being.

CHAPTER 9

HYGIENIC PORK PRODUCTION

Dr. Bijoy Kumar Sarkar

9.1. INTRODUCTION

Pigs are highly prolific animal producing two litters per year with 8-12 piglets in each farrowing. Pigs grow fast and have capacity to convert inedible feeds, forages, grain by-products, meat by-products, damaged feed and garbage into valuable nutritious meat named as pork. Rearing of pig is traditionally maintained by the economically weaker section of society, provides livelihood to millions of landless people as backyard enterprise. Pig farming demands a small initial investment on building and equipment, proper feeding, minimal maintenance and sound disease control programme to make it profitable one. Thus, pigs are ideal animal for intensive farming due to their fecundity, early maturity and short generation interval.

Although India's pigs constitute only 1% of world pig population which is widely distributed in all over the country and its rearing is an important occupation of marginal to small farmers and landless labourers of the rural society especially amongst the tribal masses. The majority of pig population in India is distributed along the central and North-Eastern region of the country. Pigs are concentrated in the North-Eastern region with almost 40% of the country's total pig population. Assam became the highest pig populated state in India with 2.1 million pig in 2019.

The majority of the pig population in India is of indigenous breeds (76 percent) which include Ghungroo, Niang Megha, Ankamali, Agonda Goan, mali and TanyVo, but the population of cross-breed and exotic pigs like Hampshire, Large White York Shire, Duroc, Landrace, and Tamworth

DOI: 10.1201/9781003749844-9

are also increasing day by day. Average meat yield from indigenous breeds is around 35 Kg/animal, which is quite low in comparison to world average of around 78 Kg/animal. Pigs are generally slaughtered at 6 to 7 months of age. The dressing percentage varies 65 to 70 percent in case of indigenous breed and 70 to 75 percent in cross-breed and exotic pigs. Around 98 percent of pig population is slaughtered every year in India.

Pork is most widely consumed meat in worldwide sharing about 37 percent of total meat consumption, whereas in India it contributes 9 percent of total meat consumption. Per capita pork consumption in India is negligible and concentrated particularly in north-eastern states including Assam, Nagaland, Arunachal Pradesh, Manipur, Meghalaya, Mizoram, Sikkim, and Tripura. Other Indian states with high pork consumption include Bihar, Jharkhand, West Bengal, Goa and Kerala. India's pork consumption can be divided into two segments. First segment being the consumption in the form of fresh pork meat sold through unorganized wet markets and meat vendors. The second segment is the high value imported pork products like sausages, ham, bacon, salami, canned meat products and frozen meat.

9.2. PORK PROCESSING PLANTS

The activities of pork processing mainly divided into three stages - slaughtering, meat cutting and further processing. Each stage involves completely different but dependent and interrelated technical operations. Significant interactions are present in between the stages and inadequacy at one stage can cause a serious negative impact (technological, biochemical or microbiological) on the product or process in a subsequent stage. Adherence to the basic guidelines in slaughtering, meat cutting and further processing and the respective hygienic regulations would contribute to the production and consumption of safe and good-quality pork and pork products and are equally applicable to both the small-scale and the medium-sized pork processing plants. Inappropriate slaughtering techniques such as defective stunning, sticking, scalding, singeing, evisceration and carcass splitting can diminish the quality of the carcass and by-products and make them unsuitable for further use. Poor standards of hygiene during slaughtering or carcass handling result in high levels of microbial contamination in the meat, thus reducing the quality and shelf-life of pork products.

All the operations in pork processing plants should be carried out in clean areas as much as possible and the products should be protected from contamination from all sources. The operational facilities should be so designed and located that a clear demarcation exists between clean and unclean sections. The following requirements are considered essential for good sanitary preparation of pork and pork products.

Floors: All the floors should be made up of brick, tile, smooth concrete or other impervious, easily cleanable, non-slippery, waterproof materials. Floors should be sloped toward the drains with a gradient of about 1:50.

Drains: To carry away waste liquids, there should be efficient drains of proper size that are correctly located, trapped and vented. Generally, for adequate waste disposal, one drain is needed for each 18 m^2 of floor space in slaughtering areas, and one drain for each 46m^2 in processing and other areas.

Walls: Glazed tile, smooth cement plaster, rustproof metal panels and properly caulked smooth plastic panels are acceptable for walls because of their non-absorbable, light reflection, effectively cleaning and sanitizable property. Ceiling should be at a height of at least 5 meters and must be tight, smooth, moisture-resistant and free from any scaling. All light bulbs should be covered with unbreakable material to prevent falling of broken pieces into the product.

Doors: All doorways through which the product passes, whether suspended on rails or lying on hand trucks, should be wide enough to ensure that the meats never touch the doorways avoid the risk of contamination. Wooden doors and doorways should be metal covered with tightly soldered seams.

Ventilation: Adequate ventilation system is mandatory to prevent excessive heat and stem condensation. Indeed, ventilation is provided to prevent accumulation of odour and dust. Window netting is essential to prevent the pest infestation and at the same time maintaining the required air flow.

Water supply: Abundant potable cold and hot water must be distributed to all parts of the operation to cope with both clean and unclean operation in the abattoir. Around 450 litres of water is required to slaughter each pig.

Lighting: The light intensity of 550 lux is required in the areas where the critical examination of sanitary control or cleanliness is conducted. In processing areas, light intensity of 220 lux should be provided for adequate visibility. In all other areas, such as dry storage, 110 lux of light intensity should be provided for smooth operation and maintaining the sanitary standards.

Refrigeration: The main purpose of refrigeration is to cool the meat after slaughter and to maintain it in a chilled state for shorter or longer storage periods required for further processing. For meat freezing, the lowest possible temperature is advised for maximum shelf-life with a recommended temperature of -18°C for freeze storage. All pork processing rooms, except in the slaughter and cooler storage areas, a temperature of about 12°C should be maintained.

Equipment: All the equipments should be made of stainless steel or plastic, rust resistant and easily cleaned and sanitized. Any stationary equipment must be located far enough from walls for the ease of proper cleaning around and under it. Foot-pedal or knee-operated wash-basins of hot and cold water, soap and disposable towels should be conveniently located in all areas. In slaughtering areas, lavatories should be facilitated with worker's dressing room. Hot-water (82°C) containers should be available for sanitizing contaminated tools. Rails must be located high enough (Bleeding rail- 2.4 metres; Dressing rail – 2.25 meters; Cooler rail – 2 meters) to prevent carcasses from touching the floor and far enough from fixed objects and walls to avoid contact.

Personnel Hygiene: Personnel hygiene with clean hands, clothing and good hygienic practices are absolutely essential for the production of good quality pork products. Workers must wear clean and protective clothing. White or light-coloured coats or gowns should be used to cover the street clothing while handling exposed product. Garments that become soiled or contaminated should be changed when necessary. All persons working with exposed meat products should have their hair under control, either completely covered with a clean cap or hat or confined by a hairnet to prevent hair from falling into products. Safety devices such as aprons, wrist guards and mesh gloves must be made of impervious material, clean and in good repair. Light-coloured clean rubber or plastic gloves should be worn by product handlers. Wearing of any kind of jewellery, badges or buttons are not allowed for person working in pork processing industry as these may accidently included in the product. Cleaned and waterproof shoes and boots made of impervious materials should be worn at all times and should be appropriate for the operations being conducted. Workers should not smoke or use tobacco in areas where edible products and ingredients are handled, prepared or stored, or where equipment and utensils are cleaned. Scratching the head, placing fingers in or around the nose or mouth, sneezing or coughing is not allowed during handling of edible products. Hands should be thoroughly washed and cleaned after using the toilet, smoking, coughing or sneezing, handling money, garbage and soiled or infected material.

Routine Cleaning: The floors should be kept clear of all debris and must be frequently washed down. A thorough cleaning programme should be followed at the end of each day. All matter should be removed from floors, platforms, gullies, etc., followed by a thorough hosing down of walls, floors and all surfaces to loosen dirt. Finally, a strong cleaning solution should be applied and left for a while before being rinsed off. A thorough inspection should be made afterwards and any areas remaining soiled should be cleaned again. A standard cleaning routine of the equipment should be established to maintain the cleanest possible products. Initially all large pieces of refuse material should be scraped or swept together and disposed off. Follow-up should include scrubbing of the equipment using brushes and a soap or detergent and a complete sanitizing with hot water of 82°C and an approved chlorine or iodine rinse. Finally, a coating of light mineral oil can be applied to metal equipment.

9.3. PIG SLAUGHTER EQUIPMENTS

Possibly, all equipment should be made of stainless steel or approved plastic, be rust resistant and easily cleaned and sanitized. Equipment which does not get in contact with the meat (overhead rails, working platforms, knocking pen) is usually made of galvanized steel.

Basic equipment needed for pig slaughter:

- Electrical stunning machine or Elther apparatus
- Sticking knive -15 cm sharpened on both sides
- A sharpening steel
- Oil or water sharpening stone
- Scabbard and belt for holding knives
- Meat saw
- Block and tackle or chain hoist
- Spreader - gambrel or metal pipe
- Bucket
- Working platforms
- Scalding barrel or tank
- Pot, barrel or system for boiling water
- Bell scrapers
- Solid scraping table or platform
- Thermometer registering up to 70°C
- Hog or hay hook
- Torch or flame for singeing
- Knock pen
- Bleeding hooks
- Blood-catching trough
- Wash trough (tripe)
- Hand wash-basin
- Implement sterilizers

All equipment coming into contact with carcasses or meat should be clean thoroughly. Knife sterilizers should be accessible to every operator. Each operator should have at least two knives, one to use while the other sterilizes.

SLAUGHTER AND DRESSING OF PIG

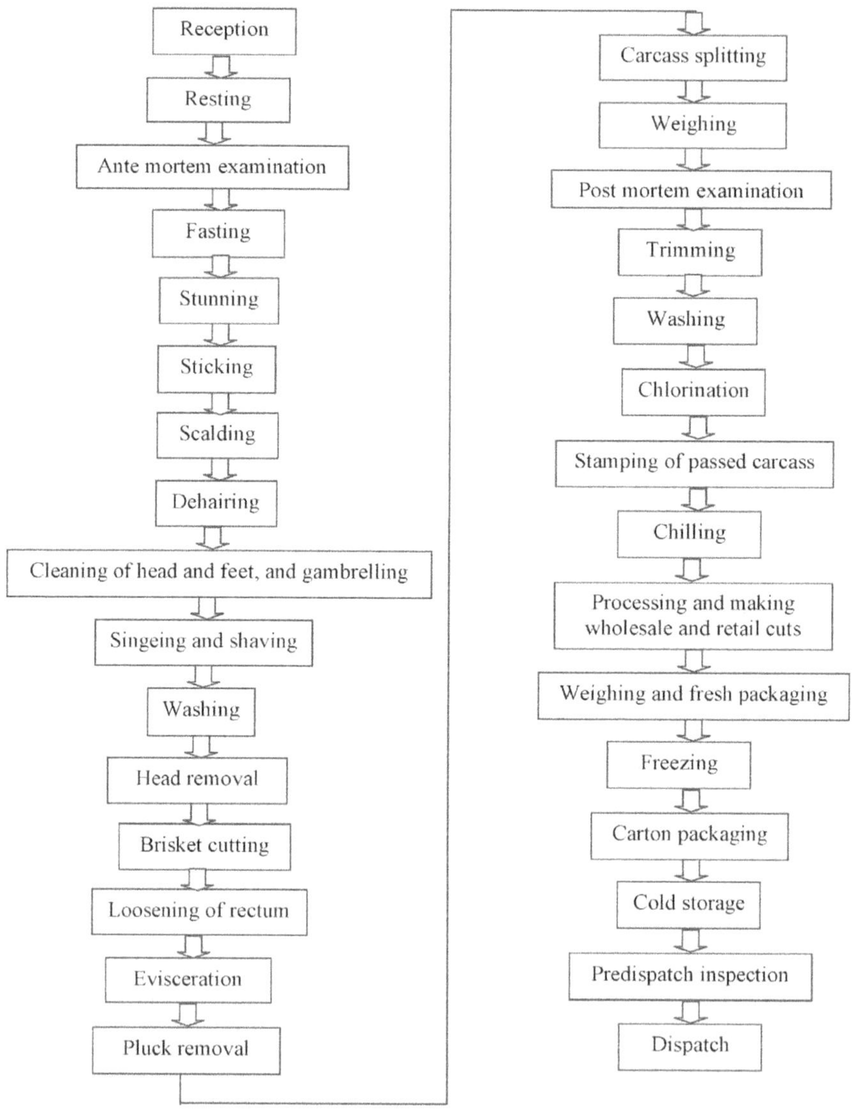

Reception: Pigs get easily excited in new surrounding especially during transport. Handling of pigs should conform to humane standards at every stage to safeguard the animal welfare as well as meat quality. According to animal welfare viewpoint, deprivation of water or food, rough handling, exhaustion due to transportation over long distances, fighting of animals etc. are stress to the animal and should be avoided to evade deleterious effects on meat quality. Even death is not uncommon among pigs during transportation in poorly ventilated, overcrowded trucks in hot weather. From loading on the farm to the stunning pen, animals must be treated kindly, and vehicles, equipment and lairage facility for livestock handling must be designed to facilitate humane treatment. It is advisable, not to mix pigs from different farms during transportation and in lairage to avoid fighting. During loading and unloading of pigs, a ramp should be used to avoid unnecessary stress to the animal.

Resting: Pigs should have a rest period after arrival at the slaughterhouse. Due to excitement, there will be loss in glycogen reserve. Generally, after a long journey, a good period of rest and adequate feed is given to pigs in holding pen to restore normal physiological condition before slaughter. It is a better practice to give molasses along with water to pigs to enhance glycogen reserves in the body. Lairage serves as resting ground for the tired and stressed pigs. The lairage should have small pens with curved corridor. Lairage should have adequate litter and drainage facility to avoid faecal soiling of skin. Depending on the length and mode of journey, age, sex and condition of pigs a minimum resting period of 2-3 hours is mandatory. Undue holding is also not advisable.

Ante Mortem Examination: Ante-mortem examination of pigs waiting for slaughter is mandatory in order to produce wholesome pork and thus safeguard the health of pork consumers. Ante-mortem examination is conducted by the qualified veterinarians with the help of para-veterinary staffs in lairage. Pigs must undergo ante-mortem examination on the day of their arrival or before the beginning of daily slaughtering. The inspection must be repeated immediately before slaughter if the animals have been held for more than 24 hours. Lairage should have proper light and ventilation for inspection and provision of isolation pen for diseased and injured animal. Only fit and healthy pigs may be slaughtered for human consumption.

Fasting: A minimum 12 hours of fasting with ad-libitum water will enhance the carcass quality. Fasting before slaughter reduces the volume of gut contents and hence bacteria and therefore reduces the risk of contamination of the carcass during dressing. In large processing plants, a mechanical shower should be provided to clean the pigs before slaughter to remove the adhered dirt and to reduce chances of microbial contamination transferring from skin to the carcass.

Stunning: One animal should not be slaughtered or stunned in front of other animals. Stunning of pigs is generally done by electrical stunning (Elther apparatus) or by chemical stunning (carbon dioxide stunning).

Elther apparatus with a pair of tongs carries electrodes by which alternating current is passed through brain. A high frequency current (250 mA) of comparatively low voltage (75 volts) is used for 10 seconds by application of tongs with strong pressure at the base of the ears on either side of the head. The electrodes should kept wet in a 20% saline solution to enable the current to pass easily through hair and skin. It produces instantaneous unconsciousness for about 30 seconds. If the current remains low, pigs become paralysed but it remains fully conscious, which is known as missed shock. On the other hand too high current may cause splash, characterized by appearance of petechial haemorrhages throughout the subcutaneous tissue due to rupture of capillaries as a result of excessive increased blood pressure.

In chemical stunning pigs are exposed to 65% carbon dioxide concentration through oval type tunnel, dip lift system or revolving wheel for 45 seconds. Chemical stunning of pig requires more space and time; and also not economical until large numbers of pigs are slaughtered at a time.

Sticking: In sticking/ bleeding of pigs the sharp end of knife with 6-8 inch-long blade is inserted in the midline if the neck at the depression in front of the sternum and then pushed forward to sever the anterior vena cava of brachiocephalic trunk at the entrance of the chest. Sometimes carotid artery is also pierced. Care should be taken not to insert the knife too far in the chest puncturing the pleura, which may cause blood aspiration in the thoracic cavity resulting adherence of blood to the parietal pleura especially the posterior edges of ribs, known as back bleeding or oversticking. A bleeding time of 5-6 minutes should be provided for complete bleeding (around 50% of total blood) and generally blood yield of pig is 2-3 kg. Pigs should be as clean as possible during stunning and sticking or bleeding.

Scalding: After complete bleeding pigs are dropped in a scalding tank (thermostatic controls and timers) maintained water at around 60-62°C for about six minutes to loosens the hair in the follicle. At low temperature hair will not be loosened and at high temperature the skin will be cooked, thus it is difficult to remove the hair. The simplest scalding equipment consists of a tank into which the pig is lowered by a hoist and water is heated by oil, gas, electricity or an open steam-pipe. Scalding water should be changed frequently to reduce contamination.

Dehairing, Cleaning of Head and Feet, and Gambrelling: The pigs should be raised on the overhead rail. Dehairing is done with a specially formed scraper (bell scraper or dull knife). If the scalding is effective all the hair can be removed by manual method. Hairs should be removed watchfully particularly from the head and feet, as it is not easily removed. Usually, the blunt end of the knife is used to scrap the bristles. It should be done along the flow of the follicles. Another simple method is to dip the pig in hot resin adhesive bath and the resin is allowed to set partially after removing from the bath. Then the hair is peeled off from the root by pulling the resin. This is less labour-intensive than scraping and produces a very clean skin. Automated scalding tank equipped with rotating rubber-tipped paddles can perform scalding and dehairing at the same time. The hair is loosened by the scalding water and thus it can be removed by the rubbing effect of the paddles against the skin.

Singeing and Shaving: Singeing removes any remaining hairs after scalding, shrinks and sets the skin, decreases the number of adhering micro-organisms and leaves an attractive clean appearance. It may be done with a hand-held gas torch or blow lamp which chars the remaining bristles and fine hairs. Automated singeing systems transport the pig into a furnace and leave it long enough for an effective singe. Thus, singeing sterilizes the cuticle and firms up rind giving it a better appearance and keeping quality. After singeing, black deposits and singed hairs are scraped off, thoroughly cleaned and the whole carcass is shaved with a fine knife to remove all the charred bristles and hairs.

Washing: The entire carcass is washed thoroughly with clean, cold running water to remove the blood stain and other dirt adhering to the carcass surface. In large processing plants, a mechanical shower should be installed to clean the carcass to reduce the chances of microbial contamination to the carcass.

Head Removal and Brisket Cutting: The head is removed before the carcass is open at the level of atlanto-occipital joint by making a circumcision around the neck approximately 2 inch behind the base of the ear. The carcass is then hoisted from the ground level to a working height by introducing a gambrel on either side of the achilles tendon in the hind limbs. A mechanical saw is used for splitting of the brisket but care must be taken not to puncture the viscera.

Loosening of Rectum: Loosening and tie off the rectum is essential to avoid further faecal contamination, which is termed as bunging. Loosening of rectum is started with loosening of the anus.

Evisceration: A mid-ventral incision is made from the mid-point of pubic symphysis up to the level of midpoint of the neck. Initially the sharp end of the knife is introduced to cut the sub-cutaneous fat. Later on, a small incision is made in pelvic cavity and by introducing the middle and index finger through the opening, the cutting end of the knife is directed forward so as to expose the pelvic and abdominal cavity. Visceral organs are pulled out except the kidneys. The pubic synthesis of the pelvis is split and anal sphincter is removed intact. Now, rectum can be detached and removed from the abdominal cavity with the entire gastrointestinal tract intact. The kidneys are usually removed after the backbone of the carcass has been split down. Leaf fat is also collected. No attempt is usually made to seal the pig oesophagus.

Pluck Removal: By using a cutting knife, the sternal cartilage is cut open to expose the thoracic cavity. The thoracic organs including the diaphragm are removed. The thoracic viscera are removed intact and care must be taken to avoid contact with the floor or standing platform. After removal of the pluck (larynx, trachea, lung, heart and liver), they are carefully examined.

Carcass Splitting: The dressed carcass is made into two equal halves by cutting across the vertebral column. The sharp end of the knife is introduced to incise the sub-cutaneous fat. The carcass is cut into two equal halves by cutting across each vertebra with the help of a cutting knife. In large processing plant, the carcasses are split along the middle line (backbone) through the skin and body wall with a saw or cleaver from the crotch to the neck. Cut is made through the pelvis and the bladder and sexual organs are removed. Sawing gives a better result but care should be taken to remove the bone dusts. If a cleaver is used, it may be necessary to saw through the rump and loin in older animals. The saw and cleaver should be sterilized in hot water (82°C) in between carcasses.

Weighing: Weighing of carcass is necessary to calculate the dressing percentage, meat yield and various cut up part percentage. Care should be taken that all damaged or contaminated parts are removed and the presentation has been standardized before weighing.

Post Mortem Examination: As soon as carcass dressing is completed, post-mortem inspection of carcasses, offal and head should be carried out by qualified veterinarians in a hygienic manner avoiding unnecessary cuts. Meat inspection knives, sterilizers for complete immersion of knives, hot and cold running water, liquid soap, towels, stamping ink and sufficient light (light intensity-540 lux) should be provided at inspection point. If signs of disease or damage are found the entire carcass and/or offal may be condemned and must not enter the food chain, but more often the meat inspector will require certain parts to be removed and destroyed especially when abscesses are present. Any instructions from the inspector to remove and destroy certain parts must be obeyed. Post-mortem examination is very important to control the diseases and to supply wholesome meat to the pork consumers.

Trimming, Washing and Chlorination: Removing visible contamination by trimming and washing are the best practical and effective method for reducing microbial contamination. Frequent sterilization of knives and other tools used in trimming process is essential to reduce or minimise bacterial contamination. Spray washing of the carcass with clean water is essential to remove visible dirt and blood stains before chilling. Soiled carcasses should be sprayed immediately after dressing before the soiling material dries, thus minimizing the time for bacterial growth. For sanitation of carcass chlorination is done with a recommended chlorine concentration of 100 ppm. Generally, hypochlorous acid is used for chlorination of carcass as it shows better efficiency.

Stamping of Passed Carcass: Carcasses and offal passed for human consumption are applied with "Government Inspected" stamp with stamping ink solution containing 1 to 2% of fuchsine in acetic acid. Only stamped carcasses should be allowed to sell in market for better consumer safety.

Chilling: Carcasses should go into the cooler as soon as possible and should be as dry as possible to retard bacterial growth and extend the shelf-life. Carcasses must be placed by hanging on rails after weighing. The height of the rails in chilling room should be 2 meter. Core temperature (temperature in deep inside the carcass) of carcass is very important and must be measured with a probe thermometer. As a general guide a deep muscle temperature of 6 to 7°C should be achieved within 12 to 16 hours, failure to which will result rapid multiplication of bacteria deep inside the meat leading to off-odours and bone-taint. In the chilling room, relative humidity of about 90 percent with an air speed of about 0.5 m/second is maintained. The cooler should not be overloaded beyond the maximum load specified by the manufacturers and spaces should be left between carcasses for the cold air to circulate. In the chilling room 0.7 meter rail spacing and 0.3 to 0.4 meter space between the carcasses should be maintained for adequate cold air circulation. Once filled, a cooler should be closed and when emptied, it should be thoroughly washed before refilling.

Processing and Making Wholesale and Retail Cuts: The primal cuts of pork are ham, loin, bacon, boston butt, picnic and jowl. Ham and loin are the highest priced cuts of a pork carcass. Boston butt and picnic, the top and bottom portion of shoulder respectively are also highly priced. Wholesale cuts are further cuts into various retail cuts as per demand of retail market.

Weighing and Fresh Packaging: Simple packaging of fresh meat with plastic foil has become very popular with the availability of suitable and inexpensive film. The main objective of simple packaging is to provide hygienically protected portioned meat for self-service retail outlets and to reduce further contamination from airborne micro-organisms. Simple packaging for retail sale in self-service outlets usually involves placing the meat portion in a plastic tray and overwrapping with a clear plastic film. Proper packaging is required with plastics, aluminum foil, freezer paper or freezer bags prior to freezing the pork. For prolonged frozen storage of pork cuts, moisture-proof, airtight packaging is mandatory to prevent freezer burn. Before sealing, air must be removed as much as possible from freezer bags. Vacuum packaging can be a good choice for packaging prior to freezing.

Freezing: Freezing is the method of choice for long term preservation of pork cuts. It stops the microbial growth and retards the action of enzyme. Fresh pork should be wrapped properly with suitable packaging film to prevent the freezer burn. Fast freezing of pork is preferable over slow freezing to overcome the problems like muscle fibre shrinkage, distorted appearance and high drip loss. Pork carcasses can be stored at -30° C to -34° C (core temperature $< -18^{o}C$) for 4 months by freezing.

Carton Packaging: Carton packaging is mandatory for bulk packaging of pork as it helps in effective utilization of transit space. Such packaging should meet all the necessary requirements for good transport worthiness. For carton packaging, carton box should be strong, compact and capable to retain its maximum strength against drop, motion, vibration and stacking load during transportation even under humid conditions. Fibre-based board has properties to make it ideal for freezing. Solid boards devoid of insulating air bubble trapped in it are preferred for carton packaging as it speeds up the freezing process of the product it contains. Carton packaging should be lined inside with polyethylene to meets the hygiene and food safety requirements.

Cold Storage: Packaged pork cut should be maintained at low temperature (-21° C to -22° C) during transportation and until further processing.

Pre-Dispatch Inspection: Packaged pork should be inspecting before dispatch by doing various tests like drop test, inclined impact test, vibration test, stack load test, rolling test etc. Label of the packages bearing required information regarding products should also be inspected carefully.

Dispatch: After proper pre-dispatch inspection, only worth packages should be dispatch under adequate low temperature until further processing.

9.4. PORK CUTTING

Halving of pig carcass is done immediately after dressing of animal and attempt should be made to saw the carcass into equal sides (halves) through the centre of the backbone. The side to be cut should be laid on the cutting table with the inside up. The primal cuts of pork are: ham, fore-end (Boston butt and picnic), loin and belly.

Pork Cutting Equipment

- Solid cutting table made of non-corrosive stainless steel with hard plastic top
- Sharpening stone
- Boning knives - 20 cm straight
- Steak knives - 30 cm curved
- Meat saw
- Meat trucks and bins
- Wrapping table
- Paper or plastic foil/bags for meat wrapping
- Tool holder
- Mask
- Gloves
- Aprons
- Hand wash-basin
- Knife sterilizer

Hind-foot: Hind-foot is removed through the hock joint at a right angle to the long axis of the leg by sawing.

Ham: Ham may be removed to make either long-cut or short-cut hams. To make short-cut hams, sawing is done between the second and third sacral vertebrae perpendicularly to the long axis of the ham. After severing the bone, knife is used to complete the removal of the ham. The ham is further trimmed by by removing the tail bone on one side and the flank on the other side. In order to obtain a long-cut ham the division is made between fifth and sixth lumbar vertebrae. The long cut is composed of a rump or chump portion and a leg portion comprising of centre section and shank portion.

Fore-foot: Fore-foot is removed by sawing through the junction between the fore-shank and the fore-foot bone at a right angle to the length of the foot.

Fore-end: Fore-end is removed by sawing perpendicularly to the length of the backbone between the third and fourth ribs from the head end. The jowl is removed by a straight cut parallel to the cut that separates the fore-end from the side just behind the site where the ear was removed. The fore-end may be divided into two cuts (boston butt and picnic) by sawing just below the exposed lower end of the blade-bone parallel to the top of the shoulder.

Loin: The middle or centre section of the pork carcass is divided into loin and belly by a straight cut from the edge of the tenderloin muscle on the ham end through a point on the front rib tight against the protruding edge of the split backbone.

Belly: Spare-ribs are separated from the belly by cutting closely underneath the ribs. Bacon is prepared from the belly by removing any thin or ragged pieces of lean. Belly is turned over and removes the lower edge with a straight cut just inside of the teat line.

The retail cuts of pork carcass are enlisted below.

a) Ham

- o Half ham (butt end)
- o Half ham (shank end)
- o Ham butt slice
- o Centre ham slice
- o Fresh ham (roast)
- o Rolled fresh ham (roast)

b) Loin

- o Boneless loin (roast)
- o Tender loin (pork)
- o Canadian style (bacon)
- o Loin chop
- o Sir loin
- o Blade loin
- o Loin (centre cut)
- o Crown pork (roast)
- o Rib chop
- o French rib chop
- o Butterfly chop

c) Boston butt

- o Smoked shoulder butt
- o Blade chop
- o Boston butt
- o Rolled boston butt

d) Picnic

- o Fresh picnic shoulder
- o Smoked picnic shoulder
- o Cushion picnic shoulder
- o Rolled picnic shoulder
- o Fresh shoulder hock
- o Arm pork steak

e) Jowl

- o Jowl bacon square

f) Belly

- o Bacon
- o Salt pork

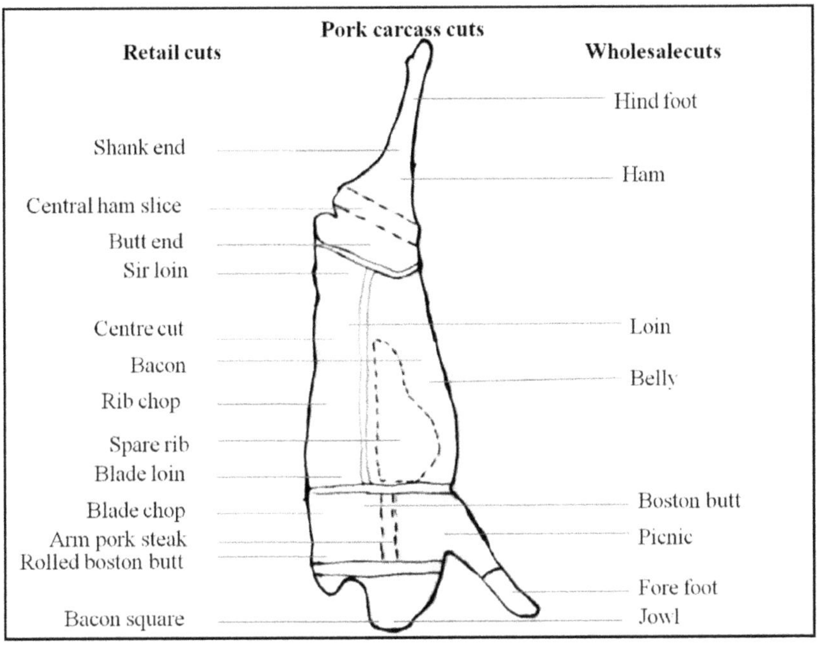
Pork carcass cuts
Retail cuts
Wholesalecuts
Hind foot
Shank end
Ham
Central ham slice
Butt end
Sir loin
Centre cut
Loin
Bacon
Belly
Rib chop
Spare rib
Blade loin
Boston butt
Blade chop
Arm pork steak
Picnic
Rolled boston butt
Fore foot
Bacon square
Jowl

CHAPTER 10

VALUE ADDED PORK PRODUCTS

Dr. Bijoy Kumar Sarkar

10.1. INTRODUCTION

Meat processing involves all the processes which modify meat either to improve its taste, flavor, nutritive value or to extend its shelf life. These include curing, smoking, canning, cooking, freezing, dehydration, production of intermediate moisture products and also the use of certain additives such as chemicals and enzymes. Meat processing excludes cutting, grinding, mixing and packaging of fresh meat in retail stores and at homes. The purpose of meat processing involves-

- Improvement of shelf life, flavor and nutritive value of meat
- Modification or up gradation of less desirable meat cuts to desirable products
- Providing convenience and variety of products

10.2. PROCESSING TECHNIQUES

Various processing techniques are available for preparation of different kind of pork products available in the market, which includes-

Emulsion Product: An emulsion is a stable mixture of two immiscible substances. Lean muscle tissue, fat, water, salt, spices, binder and extender are mixed together and subjected to high speed cutting and shearing action to prevent moisture and fat loss during cooking. It increases cooking yield and improves eating qualities particularly juiciness, texture and flavour. Preparation of emulsion products is particularly effective in economic utilization of spent animals.

DOI: 10.1201/9781003749844-10

Restructured Product: Restructured products are generally prepared from less expensive cuts, tough cuts, meat trimmings or combination of these. Small pieces of meat or meat trimmings are joined together to get bigger pieces which can be re-cut into smaller pieces to have meat pieces/ products of desirable shapes and sizes. Restructured products are processed by utilizing tumbler machine, flaking machine or tearing machine. Restructured products ensure effective utilization of low value meat cuts and trimmings with control of nutrient content and improvement in flavour, juiciness and texture. At the same time, discoloration, rapid oxidation, off flavor and presence of excess connective tissue are the common problems generally noticed in such product.

Fermented Product: Fermented product relies on microbial fermentation and dehydration to develop their specified flavour and texture. Fermented products are characterized by low pH, characteristic flavour, taste and texture and very long shelf life. Starter culture (*Lactobacillus, Pediococcus, Lactococcus and Micrococcus*) is employed in meat mix at specific temperature and humidity for a specified period to utilize sugar and produce lactic acid, causing lowering of pH of the product.

Enrobed Products: Enrobed products are also called as battered and breaded meat products, prepared by coating of meat surface with edible materials. Enrobing of meat improves appearance, crispiness, juiciness, flavor, tenderness, cooking yield and shelf life of the product.

Canned Products: Canned meat products are prepared by thermal process that employ heat (steam) to sterilize the product placed in a sterilize container (can). Retort machine is used for heat treatment. In aseptic canning, meat products and cans are sterilized separately and then sterilized meat products are filled into sterilized cans under sterile atmospheric conditions. Multiple layered flexible pouches are used in retort canning instead of metal cans. Canned products have a shelf life of at least two years at ambient temperature.

Intermediate Moisture Products: Intermediate moisture products are partially dehydrated products having a suitable concentration of dissolved solids which bind remaining water sufficiently to inhibit growth of bacteria, molds and yeasts. Lowering of water activity by addition of humectants (glycerol, sucrose, glucose and salt), retarding microbial growth by addition of anti-microbial agents (propylene glycol, sorbic acid), water activity ranging from 0.6 to 0.85 and moisture content of 30 to 50% make this product generally shelf stable without refrigeration or thermal processing.

Cured Products: Preservation of meat by salting is an age old practice in absence of refrigeration facilities. These days cured meat products are prepared by adding sodium chloride, sodium nitrite, sodium nitrate and sugar, mainly for specific flavour and colour (pink coloured nitroso haemochromogen) development of the product and also for preservative effects of curing ingredients. Cured flavour develops due to reaction between fatty acids and sodium nitrite resulting in the formation of benzonitrile and phenylacetonitrile. However, nitrite has been found to be involved in the formation of carcinogenic substance (nitrosamine) and after considering this, nitrate and nitrite levels are permitted to 500 ppm and 200 ppm respectively, as preservatives by inhibiting the growth of number of bacteria including *Clostridium botulinum.*

Smoked Products: Smoked meat products are popular and advantageous due to its characteristic flavour, colour, antimicrobial properties, antioxidant properties and surface dehydration. During smoke production by combustion of saw dust or hard wood in smoke house, undesirable polycyclic hydrocarbons are also released with desirable compounds like phenols, aldehydes, alcohols, carbonyls and organic acids. These days liquid smoke is generally used for smoking of meat, which is prepared from hard wood wherein polycyclic hydrocarbons are removed by filtration. Application of liquid smoke on the product surface imparts its smoky flavor, which is much preferred by the consumer.

10.3. COOKING METHODS

Various cooking procedures are applied for processing of various retail cuts of pork depending on its natural tenderness. Tender cuts are generally cooked with dry heat (broiling, roasting, pan broiling), whereas less tender cuts are cooked with moist heat (pressure cooking, simmering, stewing). Temperature control and time are important in pork cookery, as moisture, fat and soluble proteins are lost during cooking. Cooking losses can be minimized by controlling the cooking temperature and the final internal temperature of the pork. Time and temperature guides can be used to ascertain doneness. Cooking time is affected by the shape and size of the cut and its fat, bone and moisture content. The basic cooking methods followed for pork are mentioned below.

Broiling: Broiling of pork is specially recommended for all tender cuts. After setting the oven, place thin cuts of pork on a rack at a distance from the heat equal to two times the thickness of the cut plus 2.5 cm. On completion of one half of the desired cooking time turning of pork cuts is mandatory.

Pan-broiling: Pan-broiling is also recommended for tender cuts. For best results cook slowly over moderate heat by placing pork cuts in a hot frying-pan without adding fat or water. Cooking is continued with occasionally turning till both side of pork becomes brown and removes fat as it accumulates.

Roasting: Roasting is recommended for large tender cuts. After seasoning with salt and pepper, pork are placed by fat side up, on a rack in an open shallow roasting-pan without cover and water addition. Roast at oven temperature of 176°C to achieve desired internal temperature of 70°C.

Pan-frying: Pan-frying is usually recommended for tender cuts of up to 2.5 cm thickness. For best results cook slowly over moderate heat by placing pork cuts in a hot frying-pan containing vegetable fat or oil. Cooking is continued with occasional turning till both side of pork becomes brown and remove extra fat before serving.

Pressure Cooking: Pressure cooking is recommended for tough cuts. Higher cooking temperature can be achieved in pressure cooking which facilitates the tenderization of meat.

Stewing: Stewing is also recommended for tough cuts. In stewing meat pieces are first browned in a pan containing small amount of oil and then covered with a lid after addition of water along with curry stuff and allowed to cook at 70°C temperature in covered condition.

Simmering: Simmering is a kind of wet heat cooking, where, small amount of pork is first added to a large volume of hot water (70°C) in a tightly covered container and cooked slowly below the boiling point for the preparation of soups. The, vegetables, grains or pasta products may be added as per consumers' preference.

Braising: This method is best used for less tender cuts of pork, utilizes both dry heat as well as moist heat for proper processing of meat. After seasoning with salt, pepper, herbs and spices, pork is placed in a heavy pan. Small amount of water is added and cooking is done slowly over low heat on a stove burner under covered condition until pork becomes tender. Braising of large cuts is called pot-roasting and braising of thin cuts is called swissing.

Microwave Cooking: It is the recent development of cooking. Microwaves are high frequency non-ionizing electromagnetic waves, generated by magnetron vacuum tube within the oven. The waves penetrate the food from all directions up to a depth of 2-4 cm, causing food molecules to vibrate at high speed. Friction generated by vibration of molecules produces heat for cooking. During microwave cooking frequent turning of meat should be ensured and browning of meat does not take place in microwave cooking.

10.4.PORK PRODUCTS

Various pork products are available in worldwide. The basic processing techniques used in pork processing are curing, emulsifying, smoking and fermentation. Some popular pork products are included below.

Bacon: Bacon is prepared from belly portion of pork by curing and smoking. Brine solution is injected into the pork belly. After proper curing, bacon may be smoked or liquid smoke may be injected. Bacon is having very high in fat content mainly saturated fat and sodium.

a) **Canadian bacon**: Canadian bacon is a leaner alternative to regular bacon, which is smoked and cured pork loin. Although Canadian bacon resembles to ham in appearance and taste, but it is much similarly used as bacon.

b) **Pancetta**: Pancetta is Italian-style bacon prepared from pork belly meat by salt curing and spicing. It is often cut into small cubes and used as flavor enhancer to other foods. Sometime it is sliced paper-thin and served as a cold cut.

Ham: True ham is prepared from pork leg that has been cured and sometimes smoked. There are various types of ham in the market. Most hams are brine cured, in which a solution of water, salt, sodium nitrite, and sugar is injected in the pork leg. Some hams are dry-cured in which the meat is rubbed with salt, sugar, sodium nitrate, sodium nitrite and seasonings. These may also be called country hams. The dry-curing process draws out moisture and intensifies the colour and flavour of the pork. Unlike fresh pork, cured ham retains its rosy colour even after cooking. This is because the nitrites used in curing the pork react chemically with myoglobin to form nitroso-myoglobin, which further transform to stable pink pigment (nitroso haemocromogen), after cooking. Various types of ham found in the market include:

a) **Bone-In Hams**: Bone-in hams contain the shank bone and are available whole or in sections.

b) **Semi-Boneless Hams**: In semi-boneless hams, the shank bone is removed, leaving the round leg bone.

c) **Boneless Hams**: Boneless hams have been rolled or moulded and packed in a casing.

d) **Canned Hams**: Canned hams are brine-cured ham pieces that have been moulded, vacuum-sealed, and fully cooked.

e) **Picnic Ham:** Picnic ham is a cured-pork product, which is not a true ham, as it comes from the shoulder.

There are many ways of curing and smoking hams, each producing different flavours. Here are some of the different types of cured ham:

a) **Black Forest Ham:** This is a German ham, smoked over pinewood. It is often dipped in beef blood to produce its characteristic black surface.

b) **Westphalian Ham:** This ham is popular in Germany's Westphalian forest, prepared by smoking with juniper berries and beechwood. Westphalian ham is very dark colour, similar to Black Forest ham.

c) **Prosciutto:** This is golden-pink coloured Italian-style ham, which is dry-cured and air-dried, but not smoked. It is generally eaten uncooked, often thinly sliced and served with figs or melon. Parma is the most famous prosciutto made in Parma, Italy.

d) **Jamon Serrano:** This is Spanish dry-cured ham, prepared by using similar methods to Italian parma prosciutto. Compared to prosciutto, it has a sweet, earthy, less-salty flavour and a bit drier and smoother texture. Like prosciutto, serrano ham is also usually eaten in wafer-thin slices.

e) **Smithfield Ham:** As per state law specification, Smithfield ham must originate within the city of Smithfield, Virginia. Until 1966, this ham had to be prepared from pigs fed only peanuts, but that regulation was dropped because of the cost of raising peanut-fed hogs.

f) **Southampton Ham**: This refers to hams produced in Southampton County, Virginia. Southamptons are traditionally short shanked and a bit milder than Smithfields.

Sausage: Sausage may be defined as a meat product prepared by mincing, seasoning and forming in to cylindrical shape by natural and synthetic casings. The most familiar varieties of sausages are pork-based though it can be made from any kind of chopped or ground meat, including poultry, fish, and even some plant proteins. There are four types of sausage.

a) **Fresh sausage:** It is prepared from raw meat mix but sometimes grains (rice) or bread crumbs may be added. Fresh sausage must be thoroughly cooked before eating. The shelf life of fresh sausage is 1-2 days under refrigeration.

b) **Semi-Dry Sausage:** It is a smoked and partially dried sausage. The shelf life of semi-dry sausage is two to three weeks.

c) **Dry Sausage:** It is a fully dried sausage where smoking is optional. The shelf life of dry sausage is four to six weeks.

d) **Cooked Sausage:** It is a type of ready to eat sausage, but hot serving is preferred. The shelf life of cooked sausage is one week.

Salami: It is a fermented product usually prepared from pork. The maturation process of salami production starts after the meat is grounded or chopped and all the ingredients (salt, sugar, nitrate, spices, pepper and yeast or lactic acid bacterial starter culture) have been added and mixed properly. The maturation process is subdivided into three distinct phases-curing, incubation/ fermentation and drying/ ageing which demands specific conditions of temperature and relative humidity for each phase. Processing of salami requires three weeks or more depending upon the type of product.

Luncheon Meat: It is a canned product usually prepared from pork along with some cereal components. This product should contain at least 80% pork but the final product should not contain more than 30% of fat. Added water and cereal ingredients should be limited to 3% and 7% respectively. A batter is prepared with lean pork, pork fat, salt, nitrate, condiments, spices and ice flakes and filled in cans and sealed under vacuum. The cans should undergo for commercial sterilization at 121°C for 75 minutes. This product is stable at ambient temperature for a period of two years.

Patties: Meat patty is a popular emulsion based meat products prepared from various type of meat and generally used as filling for burger, roll or sandwich. A batter is prepared with lean meat, fat, salt, binder, extender, condiments, spices and ice flakes. The batter weighing 80-100 gm is moulded into 70-80 mm diameter and 15-20 mm thick patties, which is broiled in a preheated oven at 190°C for 20 minutes for consumption.

Loaves: This is an important ready-to-eat comminuted meat products prepared from coarse ground meat or meat emulsion or combination of both. The meat mix or batter is tightly filled in rectangular, cubical or cylindrical shaped aluminium or steel loaf pans. The pan-in mix is cooked in hot water at 80°C or in steam without pressure or broiled in hot air oven at 165°C for 3 hours to achieve the internal temperature of 70°C.

Nuggets: Meat nugget is a ready-to-eat convenient product prepared from various kind of meat including pork, obtained by cutting cooked and cooled rectangular or cubical shaped meat loaves into approximately 4 cm X 1.5 cm X 1.5 cm pieces. Usually nuggets are shallow fat-fried before consumption.

Balls: Meat ball is familiar as meat kofta to Indian consumers, which is prepared from various kind of ground meat, mixed with fat, bread powder, salt, spices and condiments in an electrically operated mixer. The dough portions of 15-20 gm are rolled into balls manually or mechanically. Balls are deep fried in refined vegetable oil at 135° C for 3 minutes to get brown color and fried flavor. Alternatively, balls can be cooked in hot water at 80°C for 10 minutes to achieve the internal temperature of 70° C, which are further light fried to get the golden brown color.

10.5. INDIAN PORK DISHES

Pork is one of the most commonly consumed meats worldwide. In India, there are many communities that traditionally do not consume pork, although pork consumption has been increased over time. People of the north-eastern regions, some parts of the south-western coast and certain communities of Kerala are major pork consumer of India. Pork dishes vary state wise with distinct traditional touches. Some of the Indian traditional pork dishes are enlisted below.

Pork Solantulem or Amsol (Goa): Pork Solantulem or Amsol is a pork dish of Goa, which is simply cooked with garlic, ginger, onion and spices. This dish is prepared with dried kokum (Garcinia indica) fruit that adds a sweet-tangy flavour with a pinkish colour to the preparation. Although, this dish is generally prepared with gravy but can also be made stir-fried.

Pork sorpotel (Goa): Sorpotel is a dark red and spicy pork dish, often made with the liver. It was originally a special Portuguese pork dish, popular in the Konkan coastal region of India, especially in Goa. It is a very popular and well cherished dish during Christmas and recently, has shown its presence as weddings and other festive dish in Goa. The meats/ organs are parboiled, diced and then marinated with vinegar and spice before being cooked.

Pork Vindaloo (Goa): The classic spicy pork vindaloo is prepared with Kashmiri red chillies, cumin, turmeric, cinnamon, peppercorns, cloves, onion, garlic, ginger, vinegar and sugar. Paste of these ingredients is heated in a heavy-bottomed saucepan containing oil, over medium heat until the oil separates and mixture starts to sizzle. Pork is added after proper seasoning with salt and spice paste and then cooked with occasional stirring until becomes brown. Then water is added and cooked under cover until the pork is fork tender.

Pork indad (Karnataka): This is a sweet and savoury pork dish of Mangalore, Karnataka. Pork indad is prepared by slicing the pork into thin slices (1/2 - 1 inch squares along with the fat on each piece) and boiled/ cooked in an open pan until all the water evaporates. Afterward, boiled pieces are fried in a pan containing ghee or oil on medium flame until the fat is released and the meat pieces become golden. Ground spices (cumin, chillies, peppercorns, turmeric, onion, garlic and tamarind) are added and fried for about 8-10 minutes on slow flame. Finally cinnamon, cloves, mint leaves, sugar, vinegar, salt are added and kept for 5-7 minutes on simmering to get the final product.

Coorg-Style Pandi Curry (Karnataka): *Pandi* means pork. It is a traditional Coorgi dish, made with an array of spices and special black vinegar (*kachampuli*). *Kachampuli* is prepared from a sour fruit called *kodampuli (Garcinia gummi-gutta).* The curry leaves give the dish a nice aromatic flavour. The spices used are roasted individually and then ground to get the best effect.

Pork Fry (Kerala): Pork fry is a common dish in the Christian homes of Kerela. It is an interesting dry-fried dish made with shallots, ginger and garlic. Additionally, a mixture of coriander powder, chilli powder and garam masala are also added which contribute to the flavour. On addition of diced or grated coconut it gives it a unique taste. Sometimes, it is garnished with coriander leaves which provide a nice fresh smell.

Tibetan Pork Ema Datshi: This cheese flavoured soupy pork dish tastes best when eaten with *senkong*, another Tibetan dish made of millet flour. *Ema datshi* is one of the more exotic Himalayan dishes beloved by Tibetans, who sometimes call it *churu*, or "rotten cheese" soup. The name 'ema datshi' means "chili" and "cheese" in Bhutanese language and it is considered as the national dish of Bhutan. This dish is prepared by cooking in a pot containing all required ingredients including blue cheese (Gorgonzola cheeses) with occasional stirring for about 20 minutes.

Pork Momos (North Bengal and Northeast India): Pork is the ideal meat for momos. Momos are popular in Tibet, Bhutan, north-eastern states of India and in northern regions of West Bengal. Momos are steam cooked dumplings made with minced pork meat and fat thoroughly rubbed and mixed with onions, ginger, green chillies and garlic. Momos are often served with a light clear soup made with pork stock.

Cheu (Assam): This is a pork dish of Assam, popular among peoples of Deuri community. Semi-boiled pork pieces are smeared with turmeric, red chilli and salt, which are then woven in bamboo sticks and roasted over fire (Charcoal).

Asan Adin (Assam): This pork dish is popular among Mishing community of Assam. This dish is prepared by boiling and smoke-drying of pork for two to three days. It is generally wrapped with Ekkum leaves (*Phyrnium spp*) and can be stored for 2 weeks in winter.

Eg-Adin Banum/ Khaophram (Assam): This pork dish is popular among Mishing and Bodo community of Assam. This dish is prepared by mild cooking of pork pieces followed by marination with local spices and finally roasted over fire until turns into golden brown

Eg-Adin Luktir (Assam): Eg-adin Luktir is a pork dish popular among Mishing community of Assam. This dish is prepared by mincing of dried pork into smaller pieces and mixing with dried bamboo soot, dried chilly and local herbs and spices.

Noausoum (Assam): Noausoum is a pork dish popular among Dimasa and Karbi community of Assam. Boiled rice is smeared over the boiled pork pieces and tightly packed in a sealed bamboo can and left for few days fermentation. Fermentation acts as preservative and enhance the shelf stability of this dish.

Sathu (Assam): Sathu is a pork dish popular among Kukki, Hajong and Hmar community of north cachar hills of Assam. For preparation of this dish, pieces of pork fat and half boiled pork are kept inside the *sathu* (water gourd) container and kept near the fire for 4-5 days. This product is having long shelf life, which can be further used in various recipes.

Classic Manipuri Pork Curry (Manipur): Traditionally this dish is cooked with pork belly. Common Indian spices like cumin, chilli powder, turmeric powder etc. are used in preparation of this dish. The use of thorny coriander leaves as garnish emits a strong fresh aroma and adds an interesting texture to this dish.

Wahan Mosdeng (Tripura): Wahan mosdeng is a popular dish of Tripura, which is very easy to prepare and incredibly tasty. Various ingredients like boiled pork, green chillies, ginger, onions and coriander leaves are tossed together in preparation of this dish. The speciality of this recipe is green chillies are roasted separately and then pounded to a paste along with ginger.

Wahan chakhwi (Tripura): Wahan chakhwi is oil free pork dish popular among Tripuri community of Tripura. *Wahan chakhwi* is prepared with pork by adding various ingredients like soda, ginger, green chilli, salt, rice flour, dry bamboo shoot, soaked bengal gram etc. The speciality of this recipe is addition of soda or ash prepared by burning of banana plants.

Wahan mukhwi (Tripura): Wahn mukhwi is a pork dish commonly known as pork chutney among the local people of Tripura. Wahan mukhwi is prepared by addition of a paste prepared with green chillies, tomato, tamarind and sugar with boiled pork. Afterwards chopped tomato, onion, cucumber, red chilli flakes, oil, salt and coriander leaves are also added. *The speciality of this recipe is addition of* tamarind, which provides sour taste to the product.

Wahan awanduru (Tripura): *Wahan awanduru* is a traditional pork dish of Tripura without addition of oil and turmeric. Small pork pieces are boiled properly and then different ingredients like dry fish, salt, and green chilly are added. Then, rice flour and garlic are added and mixed properly on flame and served after proper cooking. The speciality of this recipe is addition of rice flour, increases the consistency of soup.

Pork with Bamboo Shoots (Nagaland): This dish is a marvellous combination of pork and bamboo shoots, popular in Nagaland. Pork pieces are cooked on medium heat in a pressure cooker with occasional stirring until the oil is released. Cooking is continued with addition of chopped onions, peprika powder, salt, bamboo shoot and a paste of ginger, garlic and dry chilli with occasional stirring. After that water is added and pressure cooked for 4-5 whistles to prepare the final dish.

Smoked Pork Axone Pickle (Nagaland): Naga pork pickle is a very popular meat pickle which is not only famous in Nagaland but also in the entire north east India. The speciality of this dish is the combined flavour of smoky pork with the pungent ghost chillies, naga ginger, fermented soya bean and mejenga seeds (Sichuan peppercorns).

Bagjinam (Ngaland): Bagjinam is a pork dish of Nagaland mainly popular among the Sema Naga community. It is a fermented pork dish consumed as curry.

Naga Pork Curry (Nagaland): Naga Pork Curry is a popular pork dish of Nagaland prepared by simple cooking with few ingredients. In naga pork curry fermented bamboo shoot and a special red chilli is use commonly known as the ghost chilli, which is similar to the world-famous chilli bhut jolokia, primarily used in Assam. The gravy of naga pork curry is not thick, unlike a stew. The spicy red chilli and pungent fermented bamboo shoots give a unique taste to the product.

Smoked Pork (Nagaland): Smoked Pork is a very common item in most Naga households, which is simply mouth watering. It is often paired with axon or akhuni, which are fermented soya bean available in the form of powder, cakes or sometimes pickles. Traditional smoke dried naga pork are seasoned and the flavor is intensified with basic spices specially axone (fermented soyabean) and "Raja Mircha" (the hottest chilly) without any additional preservatives. Akhuni has a strong pungent smell that forms an amazing combination with the smoked meat. The dish forms a perfect blend between the spicy red chillies and the strong-smelling axon.

Pork Patot Diya (Nagaland): This is simple banana leaf-wrapped steamed pork dish of Nagaland. Pork, garlic-ginger paste, green chili, black pepper, tomato and salt are mixed together and the mixture is wrapped with banana leaves, which are further cooked on low flame for 30 minutes to prepare the final delicious pork dish.

Jadoh (Meghalaya): Jadoh is a popular item among the Khasi tribe of Meghalaya. It is simple but delicious dish which is primarily a mixture of red rice and pork. The special red rice gives unique flavour to the jadoh. It is cooked with onions, ginger, garlic, green chillies and bay leaves. A tasty fat infused blend is achieved by cooking the rice, meat and spices together. Since only few spices are used, the pork's smell is fresh.

Doh Klong (Meghalaya): Doh Klong is a pork dish popular in Meghalaya, prepared by frying small pieces of boiled pork in oil with local spices and served hot.

Doh Khleik (Meghalaya): Doh khlieh is a minced pork salad of Meghalaya, traditionally prepared by khasi people with pig's head and brain. Doh khlieh is prepared with boiled diced pork and pig's brain mixed with finely chopped onions, green chillies, ginger and seasonings.

Doh Kpu (Meghalaya): Doh kpu is grounded meat ball prepared of either pork or beef. Finely minced pork is mixed with onion, garlic, green chillies and local spices. Meat balls are rolled manually, fried in oil and simmered in boiling water to serve.

Zadoh Snam (Meghalaya): Jadoh snam is popular dish of Meghalaya, prepared by cooking of rice and pork in blood. Although, Jadoh snam looks like biryani, but the spices are very different with minimal use of oil. Fresh beef/pork blood is mixed with half cooked rice and fried in oil for 5-6 minutes with local spices. Blood should be carefully collected and chosen to prepare Jadoh snam, as it can alter the taste.

Doh Snam (Meghalaya): Doh snam is a blood sausage popular in Meghalaya. It is a form of sausage prepared by mincing pork bits and local spices, soaking the mixture in pig blood and wrapping it in a thin layer of pork fat.

Doh Pheret (Meghalaya): Doh pheret is a traditional product of Meghalya prepared with pig offal and spices which are cooked in a pot for long time and served hot.

Dohneiiong (Meghalaya): Dohneiiong is very famous pork dish of Meghalaya prepared with black sesame seeds. It has a unique flavour of black sesame seeds with a punch of pepper and gives a dark colour to the dish. Other few basic ingredients used in the dish are onion, green chillies, ginger garlic paste, turmeric powder, red chilli paste and mustard oil. The pork is first cooked in mustard oil with other ingredients. The black sesame seed is dry roasted and grinded into fine powder and added in the dish towards the end of cooking.

Doh Snier (Meghalaya): Doh snier is a traditional pork product of Meghalaya. It is prepared by stuffing pig's intestines with herbed blood, molten fat, spices and turmeric which later on moist cooked and toss-fried. Spices and turmeric are added to improve shelf-life of product.

Kargyong (Sikkim, Arunachal Pradesh and Darjeeling): Kargyong is a soft or hard brownish sausage like product prepared from yak/Beef/Pork and considered as curry or side dish. It is prepared with ginger, garlic and salt. It is made with natural casing called gyuma which is boiled first, then smoked and dried for 10 to 15 days. The finished lang kargyong can be eaten boiled or fried with onion, chillies and salt or to be used in curry.

Ngam Toongpak (Arunachal Pradesh): Ngam Toongpak is a popular pork dish of Arunachal Pradesh. It is prepared by mixing small pieces of pork with bamboo shoot and then put inside bamboo hollow container, which is kept with tightly covering. This product is having a longer shelf life.

Tambe-Akom (Arunachal Pradesh): Tambe-Akom is a traditional pork dish of Misimi tribe of Arunachal Pradesh. It is prepared by mixing small pork pieces with ginger, garlic and salt paste and then stuffed into bamboo hollow container, which is burnt in fire-coal. Generally Tambe-Akom is consumed in slice form.

Yoo-Aso (Arunachal Pradesh): Yoo-Aso is a traditional pork product of the Apatani tribe of Arunachal Pradesh, having very long shelf life of over 10 years at room temperature. The sub-cutaneous fat layer with skin of the trunk of pig carcass is removed and a herbal preparation called pila (aquous filtrate of the ash of millet and a locally grown fern) is uniformly applied over the surface as preservative.

Adin (Arunachal Pradesh): Adin is a traditional meat product of Arunachal Pradesh. It is prepared with pork, beef or mithun meat by smoke drying and in addition with fermented bamboo shoot.

Bai (Mizoram): Bai is one of the most easily available dishes in Mizoram. It is a stew, prepared with spinach, bamboo shoot, seasonal greens and vegetables, boiled with little salt. This is a healthy soup, as oil is not added in this dish. Baking soda is used in this dish.

Vawksa Rep (Mizoram): Vawksa rep is a traditional chewy pork product of Mizoram, prepared by light smoking of pork chunks. It is prepared with seasoned and smoked pork, chillies and fresh leafy greens.

Sa-um (Mizoram): Sa-um is traditional pork product of Mizoram, prepared by fermentation of pork fat. It is used extensively as cooking oil in preparation of different dishes.

Vawk Vun Kanpuah (Mizoram): Vawk vun kanpuah is a traditional pork product of Mizoram, prepared from pig skin by sun drying. Dried pig skin is then roasted or fried for consumption.

Vawk Vun Bawl (Mizoram): Vawk vun bawl is a traditional pork product of Mizoram, prepared from pig skin in the form of chutney.

Vawk Thithun (Mizoram): Vawk thithun is a traditional pork dish of Mizoram. It is a blood sausage prepared by stuffing of hygienically collected blood in cleaned pig intestine and then cooked. Sometime with blood, various internal organs of pig and lard are also stuffed.

REFERENCES

Aberle, E. D., Forrest, J. C., Gerrard, D. E. and Mills, E. W. (2012). Principles of meat processing. In Principles of meat science – 5th edition. Kendall Hunt Publishing, Dubuque, IA, USA.

Amit Sharma, Nibash Debbarma and P.S. Pramanik (2015). Pig management practices in Tripura. India, *Indian Journal of Animal Research*, 49 (6):863-865.

Arthur, G.H., Noakes, D.E., Parkinson, T.J. and England, G.C.W. (2001). Veterinary reproduction and obstetrics. 8th edn., W.B. Saunders Co.Ltd., London.

Arunasalam V, Chandrawathani P and Sivanandan S (1995). An Outbreak of Trypanosoma evansi infection in pigs. *Jurnal Veterinar Malaysia*. (1995) 7(2):71-73

Boro P, Bharali D, Sarma M, Sonowal M, Saharia J, Brahma J, Kalita MC and Thakuria J (2021). Performances of Ghungroo pigs reared under farm condition. *Journal of Entomology and Zoology Studies* 9(1): 2265-2267.

Bortolozzo, F.P. Menegat, M.B., Mellagi, A.P.G., Bernardi, M.L. and Wentz, I. (2015). New Artificial Insemination Technologies for Swine. *Reprod*ucton *in Dom*estic *Animals*., 50 (2): 80–84.

Chakurkar Eaknath B. & Amiya Ranjan Sahu & Sajan Naik & Chethan Kumar H.B. and Gokuldas P.P. (2021) Genetic evaluation of growth and reproductive performances of crossbred pigs reared under intensive system in tropical humid coastal climate. *Tropical Animal Health and Production* 53:243.

Cheah TS, Sani RA, Chandrawathani P, Sansul Bahari and Dahlan I (1999). Epidemiology of T. evansi infection in crossbred dairy cattle in Malaysia. *Tropical Animal Health &Prod.* 31:25-31 2

Chenoweth, P.J. and Lorton, S.P. (2014). Animal Andrology: Theories and Applications. CAB International, London, UK.

Dandapat A, Choudhury D, Debbarma KB, Das MK (2010). Phenotypic characterization of Mali pig in Tripura, India. *Livestock Research for Rural Development.* 22(4):1-4.

Dr. T. Sivakumar & Dr. S. Meenakshi Sundaram. Livestock Production Management (As Per Revised VCI Syllabus - 2016). Published by Dept. of LPM, Madras Veterinary College. 2018.

Evans, A.C.O., O'Doherty, J.V. (2001). Endocrine changes and management affecting puberty in gilts. *Livestock Production Science*,; 68: 1–12.

FAOSTAT Databases. (2020). Available from: URL: http:// faostat.fao.org/ . (Accessed in September 2020).

Gilbert, C.L., Murfitt, P.J.E., Boulton, M.I., Pain, J., Burne, T.H.J. (2000). Effects of prostaglandin F2á treatment on the behavior of pseudo pregnant pigs in an extensive environment. *Hormones Behavior*., 37: 229–236.

Gracey, J. F., David S. Collins and Robert J. Huey. (1999). Meat Hygiene. 10th edition, W.B. Saunders.

Hamill LC, Kaare MT, Welburn SC and Picozzi K (2013). Domestic pigs as potential reservoirs of human and animal trypanosomiasis in Northern Tanzania. *Parasites & Vectors*.6 (1), p322

https://nbagr.icar.gov.in

Hui, Y. H. (2012). Shelf-stable processed meat products. In Handbook of meat and meat processing, 2nd edition. CRC Press, Boca Raton, FL, USA.

I A Wright, R Deka, W Thorpe and M L Lapar. The pig sector in North East India: status, constraints and opportunities. Contributed paper prepared for presentation at the international symposium 'Sustainable Land Use and Rural Development in Mountainous Regions of Southeast Asia', Hanoi, 21-23 July 2010.

Indian Livestock Census. (2019). Animal Husbandry Statistics Division, Department of Animal Husbandry Dairying and Fisheries, Ministry of Agriculture, Government of India.

Introduction to Animal Reproduction Introduction to Animal Reproduction. In cooperation with the Agricultural Education Department and the College of Agriculture, Food and Natural Resources University of Missouri Columbia. https://www. missouricareereducation. org/doc/ animalrepro/ Student Resource. Pdf.

Johnson, L.A., Weitze, K.F., Fiser, P. and Maxwell, W.M.C. (2000). Storage of boar semen *Animal Reproduction Science*, 62: 143–172.

Jovicic, M., Chmelíkova, E. and Sedmikova, M. (2000). Cryopreservation of boar semen. *Czech Journal of Animal Science*, 65(4): 115–123.

Katiyar, A.K and Vegad J.L. (2021). A Textbook of Veterinary Special Pathology: Infectious Diseases of Livestock and Poultry.

Khargharia, G., Zaman, G., Laskar, S., Das, B., Aziz, A., Roychoudhury, R. and Roy, T. C. (2014) Phenotypic Characterization And Performance Studies of Niang Megha and Doom Pigs of North Eastern India. *Asian Academic Research Journal of Multidisciplinary*, 1 (27):667-676.

Mankertz A. (2008). Circoviruses. Encyclopaedia of Virology (3rd Edition). Elsevier.

Maxie M.G. (2007). Jubb, Kennedy & Palmer's Pathology of Domestic Animals. 5th Edition.

Maynard LA, Loosli JK, Hintz HF and Warner RG. (1979). Animal Nutrition.7th Edition McGraw-Hill.

McDonald P, Edwards RA, Greenhalgh JFD and Morgan CA (2009). Animal Nutrition. 6th Edition. Pearson.

Merck Veterinary Manual - Swine Section

Murasing, D.K., Talukdar, D.J., Lalrintluanga, K., Ahmed, F.A., Kayina, A and Das, S. (2020). Recent advances in preservation of boar semen in liquid state: An overview. *International Journal of Chemical Studies*, 8(2): 112-115.

Mwamba T, Pandey VS. (1977). Effect of terramycin in balantidiosis of pigs. *Ann Rech Vet.*,8(2):167-9. PMID: 596795.

N.S.R Sastry & C.K. Thomas. (2013). Livestock Production Management. Kalyani Publishers, Ludhiana – 141008.

Nibash Debbarma and Dr. Peasanta Boro. Text Book on Swine production Management. (2020). 1st Edition. INSC Publishing House (IHP), Chikkamagaluru, Karnataka.

Nilotpal Ghosh. Livestock Production Management. (2019). Published by PHI Learning Private Limited, Delhi - 110092.

Noakes, D.E., Parkinson, T.J. and England, G.C.W. (2009). Veterinary reproduction and obstetrics. 9th edition. Saunders Elsevier publication, Philadelphia.

Noakes, D.E., Parkinson, T.J., England, G.C.W. (2003). Arthur's Veterinary Reproduction and Obstetrics. 8th edition., Elsevier, Amsterdam.

Nurulaini R, Premaalatha B, Zaini CM, Adnan M, Chandrawathani P, Fazly Ann ZA, Ernie Aryanti A and Ramlan M (2013). Trypanosomiasis outbreak in Deer, Cattle, Buffaloes and Pigs in Perak. *Malaysian Journal of Veterinary Research.* 4(1).

Pearson, A. M. and Tauber, F. W. (1984). Processed Meats. Springer Dordrecht.

Ponds WG, Church DC and Pond KR. (1995). Basic Animal Nutrition and Feeding. 4th Ed. John Wiley & Sons.

Rahman, M., Phookan, A., Zaman, Galib Uz., Das, A., Akhtar, F., Hussain, J and Choudhury H (2020) Growth and Reproductive Performances of Doom Pigs Under Field Condition. *Veterinary Research International*, 08 (02): 73-77.

Rajkhowa S, Kalita C and Sarma DK (2014). Enterotoxigenic and attaching and effacing Escherichia coli strains in diarrhoeic piglets and their antibiogram. *Philippine Journal of Veterinary Medicine*, 51: 30–37.

Rajkhowa S, Neher S, Pegu SR and Sarma DK (2018). Bacterial diseases of pigs in India: a review. *Indian J. Comp. Microbiol. Immunology Infectious Disease*. 39 (Special Issue), 29-3.

Ranjhan SK. 2001. Animal Nutrition in the Tropics. Fourth Revised Edition. Vikas Publishing House Pvt. Ltd.

Rasool G (2016). BLOOD RADOSTITS Veterinary Medicine. 10th Edition.

Reddy DV. (2015). Applied Nutrition (Livestock, Poultry, Rabbits and Laboratory Animals). Third Edition. Oxford & IBH Co. Pvt. Ltd.

Roberts, S.J. (1971).Veterinary Obstetrics and Genital Diseases. 2nd edition., CBS Publisher and Distributor, New Delhi.

Rutsa, M C and Rutsa V (2019) Characterization of population and production system of Tenyi-vo pig of Nagaland (India). *Journal of Livestock Biodiversity*, 9: 5-8.

Saikumar G and Das T. (2019). Recent Advances in Animal Virology. 6.171–195.

Saxena, M.S. (2000). Veterinary Andrology and Artificial Insemination. CBS publishers & distributors Pvt. Ltd, New Delhi, India.

Senger, P.L.: Pathways to Pregnancy and Parturition. 2nd edition., Cadmus Professional Communication, Richmond, Virginia, USA. 2005.

Sharma, B. D. (2003). Modern Abattoir Practices & Animal Byproducts Technologies. 1st edition, JAYPEE Brothers, New Delhi.

Shastri ,G. A. (2020). Veterinary Pathology. CBS Publishers. India

Singh, M and Mollier, R.T. (2020). Artificial Insemination in Pig, its Status and Future Perspective in India: A review. *Indian Journal of Animal Sciences*, 90 (9): 1207–1212.

Soriano-Ubeda, C., Matas, C. and Garcia-Vazquiz, F.A. (2013). An overview of swine artificial insemination: Retrospective, current and prospective aspects. *Jounal of experimental and applied animal science*, 1(1):67-97.

Stewart, K.L., Flowers, W.L., Rampacek, G.B., Greger, D.L. (2010). Swanson, M.E., Hafs, H.D.: Endocrine, ovulatory and reproductive characteristics of sows treated with an intravaginal GnRH agonist. *Animal Reproduction Science*, 120: 112–119.

The Merk Veterinary Manual, *https://www.bing.com/ck/a*

TUR Ýrfan. (2013). General reproductive properties in pigs, *Turkish Journal of Veterinary and Animal Sciences.* 37: 1-5

Wahab A Rahman, Samatha Fong, Chandrawathani P, Nurulaini R, Zaini CM and Premaalatha B (2012). Comparative seroprevalences of bovine trypanosomiasis and anaplasmosis in five states of Malaysia. *Tropical Biomedicine.* 29(1): 65–70.

Wahab AR, Zary S, Cheah TS and Chandrawathani P (2002). An assessment of the prevalence of Trypanosoma evansi in cattle from some farms in Perak, P. Malaysia, using the QBC technique. *Journal Biosains*.13(1): 3.

Zosangpuii, J B Rajesh, H Lalliankimi, L Hmar, NS Singh and Elizabeth L Leihang. (2020) Phenotypic characterization of local pigs (Zovawk) in Mizoram. *Journal of Entomology and Zoology Studies* 8(5): 336-338.

For Product Safety Concerns and Information please contact our EU representative GPSR@taylorandfrancis.com
Taylor & Francis Verlag GmbH, Kaufingerstraße 24, 80331 München, Germany

www.ingramcontent.com/pod-product-compliance
Ingram Content Group UK Ltd.
Pitfield, Milton Keynes, MK11 3LW, UK
UKHW020316270926
475702UK00050B/751
* 9 7 8 1 0 4 1 2 4 7 1 6 6 *